Legitimation in a World at Risk

E.L. Desmond

Legitimation in a World at Risk

The Case of Genetically Modified Crops in India

E.L. Desmond
School of Sociology and Philosophy
University College Cork
Cork, Ireland

ISBN 978-981-10-6064-9   ISBN 978-981-10-6065-6   (eBook)
https://doi.org/10.1007/978-981-10-6065-6

Library of Congress Control Number: 2017949730

Cover illustration: ZUMA Press, Inc. / Alamy Stock Photo

Printed on acid-free paper

This Palgrave Macmillan imprint is published by Springer Nature
The registered company is Springer Nature Singapore Pte Ltd.
The registered company address is: 152 Beach Road, #21-01/04 Gateway East, Singapore 189721, Singapore

May you now guard science's light
Kindle it and use it right
Lest it be a flame to fall
Downward to consume us all
(The Life of Galileo, Bertolt Brecht)

Preface

My interest in genetically modified (GM) crops arose as a result of my former employment in logistics for the multinational agrochemical company, Syngenta. The latter was formed when Zeneca and Novartis merged in 2000. The merger led to Syngenta becoming one of the world's largest developers of GM technology. It also resulted in my relocation from the UK, where I had worked for Zeneca for 5 years, to Switzerland, to take up employment with the new organisation.

At the time of the merger, GM technology was still relatively new. The Flavr Savr tomato, despite its initial commercial success, had been withdrawn following public protests. The uncertainty regarding GM technology followed a number of negative research studies, including that conducted in 1996 by Dr Arpad Pusztai, which found that GM potatoes were linked to health issues.

Following the merger, the issue of GM technology became controversial within Syngenta itself. Members of senior management were guarded in their responses to questions as to the company's future intentions in relation to research in the area. The hesitancy led me to conduct my own online research. A Google search uncovered protests in Warangal, India—most notably, in the village of Nandanapuram (a pseudonym), one of the villages featured in this book. I resigned from Syngenta after 2 years in Switzerland given my growing distance from the ethos of the company, as well as my concerns regarding pesticides and GM crops. My interest in Nandanapuram, however, continued.

This book represents the outcome of this long-term curiosity, an interest which later led me to University College Cork (UCC) in Ireland, where I undertook a master's degree in Sociology. My master's thesis explored the theme of GM crops in the Irish context. This was followed by a doctorate, funded by scholarships from the Irish Research Council and the W.J. Leen Trust at UCC. This financial support allowed me to undertake the 9 months of ethnographic fieldwork in India which is central to this book.

The decision to undertake a doctorate arose from my desire to understand how the widespread adoption of GM crops at times coincided with protests against them. From my work on the theme in Ireland, where the commercial cultivation of GM crops is currently banned, I realised that the type of research I needed to undertake would require a location where GM crops had been grown for a number of years and where their adoption involved conflict.

My Internet search in Switzerland in 2002 had led me to Warangal, the district in Telangana where the research for this book is located. At the time of the fieldwork, Telangana was located in the state of Andhra Pradesh. Following an ongoing struggle for secession which reached a peak during the research period, however, Telangana became a separate state in 2014.

India is extremely relevant as a location for this research given that it is a high-risk agricultural context associated with a protracted agrarian crisis. It is also the site of well-publicised farmer suicides with purported links to Bt cotton, a GM crop. The uncertainty surrounding GM crops fits well with Ulrich Beck's concept of 'risk society', a central theoretical framework of the book. The idea of legitimation in relation to risk society emerged as a result of my analysis of the impact of power relations on the differences in the cultivation methods, and approaches to risk negotiation, adopted in the three villages involved in the analysis.

I initially visited two of the villages which feature in this research (Bantala and Orgampalle, both pseudonyms) during a 10-day pilot study undertaken in March, 2010. Nandanapuram was visited in July of the same year. While the early focus was on Bantala, all three villages were later visited on alternate weekends throughout the research period. The empirical research explores the complex trade-offs which Bt cotton entails for cultivators, both practically and politically, the assessment of which forms part of a legitimation struggle. Uniquely, this research involves extended interviews with cultivators themselves over the 9-month period between June, 2010 and March, 2011.

The book highlights the way in which the issue of GM crops is embedded within a complex set of power relations which determine the way knowledge surrounding risk is defined as both an ontological and a political concern. My research experience explores the way in which the use of technology in the areas of IT and travel has enabled the development of an interconnected world. This heightens the ability of contemporary social actors to become critically engaged globally like never before. This book is the outcome of such a globalised critical engagement.

The greatly expanded scope for reflexivity and critique in contemporary society lends itself to an increased concern with legitimation as an ongoing process of making judgements concerning the development of scientific and technological knowledge and the uses to which such knowledge is applied. The analysis examines how the legitimation of risk society involves a focus on the way in which power is exercised in terms of the distribution of resources and the recognition and representation of the risk exposure of the most vulnerable. This, it argues, is a concern not only within one's immediate locality or state but is increasingly informed by a global awareness which transcends traditional boundaries. The concern with legitimation has diffused worldwide even as exposure to risk has. This is contributing to an ongoing focus on the way in which power is exercised in relation to risk negotiation, democratic practice and knowledge construction given the increasingly interconnected attempts by humanity to negotiate a world at risk.

There are a large number of people I would like to thank for their assistance with this book of which those mentioned here represent an important, but necessarily, brief selection. They include Dr. Kathy Glavanis-Grantham (now retired) and Dr. Ger Mullally from the Department of Sociology at UCC for their invaluable support throughout my doctorate and beyond. I thank Piet Strydom, a risk theorist and a retired senior lecturer at UCC.

I am also grateful to Professor Glenn Stone from the Department of Anthropology at Washington University in St Louis for his examination of my doctoral thesis and for his own inspiring work in the Warangal district. And to my family and friends—grateful thanks for bearing with me throughout my preoccupation of recent years—especially my brothers, Tim, for his ongoing support of my writing and Don, for his calm and patient technical assistance!

In India, I wish to acknowledge the guidance of Professor Purendra Prasad at Hyderabad Central University (HCU). This has continued since my return to Ireland. I also offer thanks to the Sociology faculty in HCU, particularly Professor Vinod Jairath, the then Head of Department, and Professor Haribabu and Dr. Nagaraju for permitting me to attend their master's seminars in Sociology. These greatly enriched my research experience and drew my attention to key debates within Indian Sociology. Thanks must also go to those in the departments of Anthropology, Economics and Political Science at HCU who allowed me to attend their conferences featuring guest speakers, including the late Scarlett Epstein, Kancha Ilaiah and Shiv Visvanathan.

I thank Professor Ananta Kumar Giri for his stimulating company at conferences in Orissa and Karnataka, and for his kind invitation to present a paper on my work at the Madras Institute for Development Studies in Tamil Nadu. I also extend thanks to the International Crops Research Institute for the Semi-Arid Tropics (ICRISAT), the Center for Sustainable Agriculture (CSA), the National Institute for Rural Development (NIRD) and the Centre for Economic and Social Studies (CESS), all in Hyderabad, as well as Kakatiya University in Warangal, for the use of their library facilities. For his important help and for always being there when I got stuck, I thank Dr. Vinod Goud Vemula at ICRISAT. I am also grateful to Professor Revathi (CESS/Kakatiya University), Dr. Sharma (ICRISAT) and Dr. Venkateshwarlu (Glocal) for our highly informative discussions.

I extend thanks to Professor Aparna Rayaprol at HCU for her assistance with my accommodation at the university. I also thank the staff at Tagore International House, and my fellow residents from Scandinavia, the United States and India. You all made Tagore such a warm, welcoming place to return to between field trips. I extend sincere and grateful thanks to my translators without whom this book would have been impossible. And I thank the representatives from NGOs, politics and industry, as well as the academics and researchers, who gave of their time so willingly to talk with me. I am also grateful to Dr. Sarthak Gaurav whose highly detailed, engaged and constructive review of an earlier draft of this book added significantly to the completed work. And sincere thanks to the team at Palgrave whose professional and helpful approach made the final steps to publication such a pleasant experience.

Finally, and most importantly, I thank the villagers whose warmth, wisdom and humour will always mean so much. This book is dedicated to my parents whose love and support made it possible and to the Warangal farmers and landless labourers whose ongoing struggle to negotiate risk inspired it.

Mallow, Ireland   E.L. Desmond

Telangana State, 2017

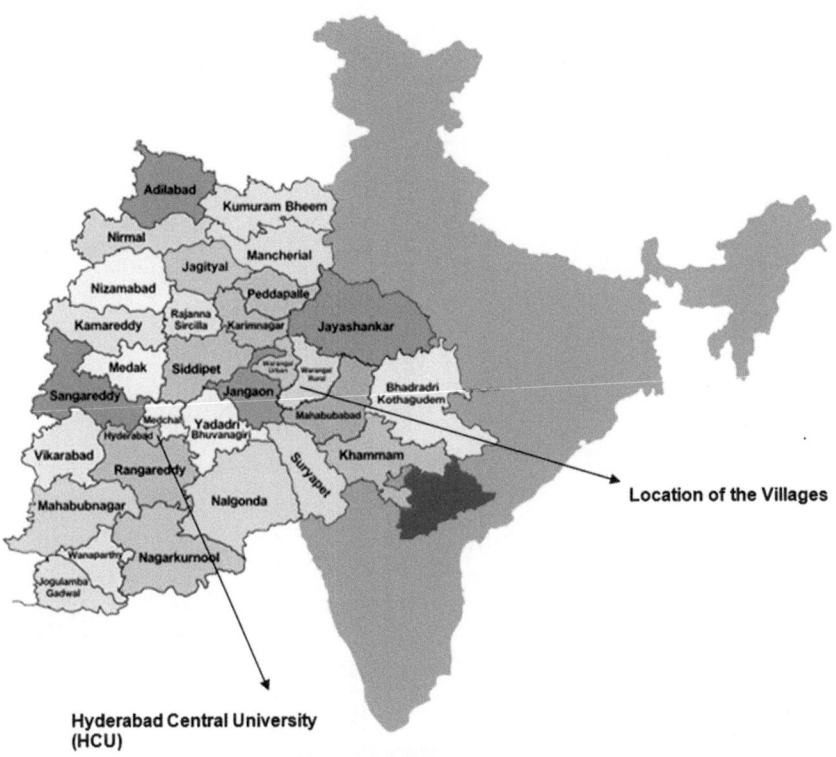

The 31 districts of Telangana State

Map 1 Telangana State, 2017
The 31 districts of Telangana State
The district of Warangal was divided into Warangal Rural and Warangal Urban in
2017 (Map reproduced with the kind permission of the Government of Telangana
Source: http://www.telangana.gov.in/About/State-Profile)

Contents

About the Author

E.L. Desmond is a sociologist whose work centres on legitimation, risk and democracy. Her book *Legitimation in a World at Risk: The Case of Genetically Modified Crops in India* is based upon 9 months of fieldwork conducted in the high-risk, politically volatile district of Warangal in the newly formed state of Telangana. Her articles have appeared in international journals, including the *Journal of Risk Research, World Development Perspectives* and *Gandhi Marg*, and a number of edited collections. Further details can be found at https://ucc-ie. academia.edu/ElaineDesmond.

Glossary

Adivasi:	India's indigenous population, also referred to as 'tribals'
Ahimsa:	Non-violence. From Sanskrit meaning 'do no harm'
Bandh:	Civic strike
Bhagavad Gita:	Seven-hundred verse scripture, which is part of the *Mahabharata* Hindu epic
Brahmin:	Also Brahman. Caste grouping at apex of caste system. Associated with priesthood.
Charkha:	Spinning wheel
Crore:	Ten million
Dal:	Thick stew made from pulses
Dalits:	Scheduled Castes. Formerly 'Untouchables'
Dharma:	Duty to be morally upright
Dharna:	Peaceful demonstration or mode of compelling debt payment by sitting at the debtor's door without eating
Dora:	Large land-holder in Telangana
Gheraos:	Encirclement of individuals or buildings by large groups as a form of intimidation in protests
Godown:	Warehouse
Goonda:	Criminal
Gram Panchayat:	Village council
Gram Sabha:	Meeting of all adult voters in the village
Hartals:	Closure of shops and offices as a form of civil disobedience

Inams:	Plots of land granted as gifts in return for services to the ruler
Jagirdar:	*De facto* ruler of a territory with right to extract revenues. Developed in the early thirteenth century during the time of Muslim rule in India
Jail bharos:	A form of civil disobedience where protestors seek arrest in order to gain recognition for their cause
Jajmani:	Occupational inter-dependence associated with the caste system
Jati:	Sub-castes or endogenous occupational groups within the caste system
Kacha house:	House made from mud with roofs of rice straw or other thatching material
Khadi:	Homespun cotton. Advocated by Gandhi in resistance to British rule
Kshatriya:	*Varna* category associated with kings, governors and warriors. Second in varna ranking after *Brahmins*
Lakh:	One hundred thousand
Lathi:	Stick which may be topped with metal. Used by Indian police as a means of crowd control
Mahabharata:	One of the major Sanskrit epics from ancient India. Estimated to date from 900 or 800 BC, it remains highly influential in contemporary India
Naxalites:	Also referred to as Maoists
Nizam-ul-Mulk:	Often abbreviated to *Nizam*. Title given to the rulers of Hyderabad State between 1724 and 1948
Pakka house:	House made from cement and bricks
Panchayat Raj Institutions:	Three-tier system of local governance involving village, mandal and district levels
Quintal:	Unit of measure for cotton yield. Equivalent to 100 kilograms
Rabi:	Rabi crops are sown in winter and harvested in spring. They include wheat and grams
Rasta roko:	A form of civil protest. Involves the blocking of roads and railways
Rupee:	India's unit of currency. At the time of the research, one euro was equivalent to between 60 and 70 rupees.

Salwar kameez:	A form of Indian dress which involves loose trousers, a long shirt and a scarf (*chunni*)
Sarf-e-khas:	Crown lands across the former State of Hyderabad which were claimed by the *Nizam*
Sarpanch:	Head person in village
Satyagraha:	A form of nonviolent resistance inspired by Mahatma Gandhi. Loosely translated from Sanskrit as 'truth force'
Swadeshi:	Self-sufficiency. Term used by Gandhi to advocate the boycotting of British goods as a form of resistance to British rule
Swaraj:	Self-governance or self-rule
Vaishya:	Third-ranked *varna* category associated with traders
Varna:	A broad category of the caste system
Vedas:	Ancient sacred texts, of which the *Rig Veda* is one of the earliest texts in the world and dates to c. 1500 BC.
Zamindar:	Revenue collector under Muslim rule and later landowner under the British
Zamindari:	System of land tenure where the *zamindar* collected rents from cultivators to pay to the government

Abbreviations

APL:	Above Poverty Line
BC:	Backward Caste
BPL:	Below Poverty Line
CPI (M-L):	Communist Party India (Marxist-Leninist)
FC:	Forward Caste
GEAC:	Genetic Engineering Approval Committee
HCU:	Hyderabad Central University. Also known as the University of Hyderabad (UoH)
IMF:	International Monetary Fund
INDIRAMMA:	Integrated Novel Development in Rural Areas and Model Municipal Areas
MFI:	Micro-Finance Institution
NGO:	Non-Governmental Organisation
NPM:	Non-Pesticide Management
NREGS:	National Rural Employment Guarantee Scheme
OBC:	Other Backward Caste
PDS:	Public Distribution System
PRIs:	Panchayat Raj Institutions
RCGM:	Review Committee on Genetic Manipulation
SHG:	Self-Help Group
SC:	Scheduled Caste
WTO:	World Trade Organisation

List of Figures

List of Figures

List of Tables

1

Introduction: Legitimation and Genetically Modified Crops in a World at Risk

This book explores the conflict surrounding genetically modified (GM) crops in three Indian villages using the concept of legitimation. It focusses on Bt cotton, currently India's only GM crop, and examines how power relations in the villages impact upon the way in which Bt cotton is variously legitimated or delegitimated by cultivators themselves. The book argues that the findings from the villages provide significant insights for the global legitimation struggle which the negotiation of a world at risk entails and within which the conflict surrounding GM crops is embedded.

GM Crops, Risk Society and the Struggle for Legitimation

The issue of GM crops in India provides an invaluable insight into the contestation regarding scientific knowledge in global society. This unrest relates to fears that the power relations which define the construction of scientific knowledge and its applications are contributing to significant ontological risk in the form of scientific innovation itself (Knorr-Cetina and Mulkay 1983; Latour 1983; Shiva 1988). The resulting uncertainty

© The Author(s) 2018
E.L. Desmond, *Legitimation in a World at Risk*,
https://doi.org/10.1007/978-981-10-6065-6_1

has led to 'a crisis of public confidence in science' (Wynne 2001: 445), a situation which, this book argues, has given rise to a central concern with legitimation in contemporary society.

Hurrelmann et al. (2007: 12–13) claim that legitimacy is a multifaceted concept. They (ibid.: 8) highlight the distinction between legitimation as a *process* of evaluation and legitimacy as the *attribute* which is being assessed [italics in the original]. This book explores legitimation in association with the concept of risk society as defined by Ulrich Beck. The theory of risk society provides the theoretical framework for the book given the congruity of Beck's exploration of risk with the uncertainty associated with GM crops explored here.

Beck (1992: 19) claims that the incorporation of science and technology into attempts to secure economic growth has meant that the 'social production of *wealth* is systematically accompanied by the social production of *risks*.' Beck (1992, 1999) argues that this has given rise to a 'risk society.' This book claims that the risks associated with wealth production are subject to a process of legitimation.

According to Beck (1992: 23), the risks produced by technological innovation in risk society are characterised by the fact that they 'induce systematic and often *irreversible* harm, generally remain *invisible*, are based on *causal interpretations*, and thus initially only exist in terms of the (scientific or anti-scientific) *knowledge* about them' [italics in the original]. GM crops are identified by Beck (2009: 74–6) as one aspect of risk society.

Within Beck's (1992: 39) theorisation of risk, he argues that the Marxist focus on class arising from the distribution of wealth in industrial society is replaced by a concern with 'risk positions' in risk society. Beck (ibid.: 23) claims that, in risk society, the relevant power relations are not the relations of production which define class but rather the individualised 'social risk positions' which differentiate 'the affected' from the 'not yet affected' (ibid.: 40).

This book argues that the categories of affected and not yet affected are differentiated according to pre-existing dimensions of stratification. These pre-existing dimensions determine the most vulnerable or those who will be affected by risk first. While Beck (1992: 35) argues that 'wealth accumulates at the top, risks at the bottom,' he does not suggest, in the absence of class, how this stratification occurs. As Chap. 2 explores, this book suggests that land-holding, caste and gender represent pre-existing dimensions of differentiation in the context of the Indian village.

The lack of inclusion of class as a determinant of social risk positions in this book does not suggest that class is not an aspect of Indian society. However, it takes the view, along with Weber ([1968], 1978: 36), that assessments of legitimacy draw from pre-capitalist conventions, such as those defined by tradition and affect. As such, the ongoing influence of caste and gender on the legitimation and differentiation of risk in the Indian context predates capitalist class relations and, in many ways, defines them.

The limited relevance of class to an analysis of Indian society is noted by Jangam (2016: 25) who argues that Marxist theory is rooted in a 'Western epistemic model' which is 'incomplete both as a theoretical and analytical category in the Indian context' (ibid.). In terms of the current analysis, it is argued that class positions are themselves legitimated in the Indian context according to pre-existing cultural beliefs surrounding caste and gender. Private ownership of land, as a means of production, can be regarded as an aspect of class in the rural context. However, it is explored here not only in terms of its economic impact on the individualised distribution of risk related to the performance of Bt cotton; it is also examined in terms of the intersection of access to land and the other means of production in agriculture with the stratification associated with caste and gender. This highlights the way in which class positions associated with access to the means of production are themselves mediated through pre-existing dimensions of caste and gender.

The analysis of legitimation and risk undertaken here, therefore, involves a blending of Weber's emphasis on pre-capitalist social markers of authority (here, as defined by gender and caste in India) with Marxist understandings of class (as defined by access to the means of production—most notably, access to land). This seeks to explore the way in which the individualised distribution and negotiation of risk is legitimated as a *social* concern.

This book argues that the dimensions of land-holding, caste and gender are central determinants not only of the way in which risk is distributed but also of the ability of particular individuals and groups to influence its epistemic construction. Giddens (2003: 22) defines risks as 'hazards that are actively assessed in relation to future possibilities.' They are, therefore, subject to probability assessments given the 'epistemic gap' (Desmond 2014: 13) and 'non-knowing' (Beck 2009: 115) with which they are associated.

The epistemic gap which risk represents means that, until risk materialises as a reality, its conceptualisation remains a social construction which is intersubjectively defined through discourse. Thus, Strydom (2002: 114) notes that '[t]he discursive construction of risk is a social process in which different social actors or collective agents compete and conflict with one another in the medium of public communication and discourse.' This book argues that both the materialisation of risk in particular contexts, and its ideological construction, are subject to legitimation struggles which are mediated through power relations at local, national and global levels.

Renn (2008: 180) claims that GM crops fall within the category of risk classified as 'ambiguous' given that they are linked to 'unresolved uncertainty' (ibid.: 179). Stone (2007: 71) argues that, not only is there a lack of knowledge in relation to the risk of the technology itself, the cultivation of GM crops is itself expanding the epistemic gap associated with such risk given the phenomenon he describes as 'agricultural deskilling.'

Within his concept of agricultural deskilling, Stone (2007: 72) claims that the indigenous knowledge associated with the treatment of pests and disease in agriculture is being replaced by a regime of pesticides and fertilisers in which cultivators have been reduced to 'passive customers of seed firms.' At a time, therefore, when detailed knowledge is crucial to ascertain the risks of Bt technology, the basis for such knowledge is being eroded through the use of the technology itself.

Stone (2011a) claims that this process of deskilling among agriculturalists is ongoing through E-agriculture where cultivators receive not only the latest weather reports and crop prices but also advice from experts on the cultivation of their crops. Such deskilling limits the ability of cultivators to bridge the epistemic gap associated with risk through reliance on their own autonomous knowledge gained from a hands-on cultivation of the land.

Stone (2007: 71) highlights that agricultural deskilling means that knowledge concerning agriculture is increasingly defined through social relations rather than environmental cues. In such a situation, the existence of powerful influencers with regard to risk negotiation becomes key. This book claims that, given this, the legitimation of influencers, and their position within the power structure of the local context, is central. It asserts that environmental learning does occur in such a context, but that this learning is mediated through power relations. Hence, building on Stone's work, this book argues that the extent to which agricultural deskilling occurs depends upon the power relations of particular contexts and the way in which power is exercised as a result of these.

In rural India, the legitimation struggle concerning GM crops takes place within an already high-risk context. This is characterised by an agrarian crisis resulting from erratic monsoons, the commercialisation and individualisation of agriculture and the reduced investment by the state in the agrarian sector, particularly with regard to irrigation (Reddy and Mishra 2010; Deshpande and Arora 2010). The declining viability of agriculture has contributed to the high-profile issue of farmer suicides. One quarter of a million farmer suicides (256,913) were recorded for India as a whole between 1995 and 2010 (Sainath 2011).

Beetham (2013: 16–17) argues that legitimacy incorporates a concern with legality, normative validity and consent. The legitimation struggle with regard to GM crops involves contestation of their validity in terms of their ability to alleviate the already high risk of the Indian context, as well as their likely impact on human health and the environment. The attempt at establishing this validity, however, has increasingly drawn attention to the power relations involved in the legitimation struggle, a focus which has itself problematised efforts to establish legality and consent.

The challenge to the validity of GM crops is undertaken globally by high-profile campaigners such as the Indian activist, Vandana Shiva. The validity of the technology is also problematised by social movements, most notably those organised through NGO activism at international, national and local levels, as well as by media debate and mass protests. Scoones (2008: 325) notes the way national protests are increasingly supported through global networks as part of a worldwide anti-GM campaign. Through these public demonstrations, opponents assert that the technology poses uncertain environmental, ecological and health risks which represent a serious threat to the future survival of humanity.

The position of activists is challenged by global multinationals, such as Monsanto, the corporation primarily associated with the research and development of GM crops in India. The marketing of the technology entails corporate assertions of its validity not only on the grounds of its safety but also of its necessity. Proponents assert that the technology's potential to increase crop yields and to enhance the nutritional value of existing food staples will be vital to addressing food shortages, particularly given the world's growing population (Prakash and Conko 2006: 41–42). The current study examines the differing impact of this contested validity on the micro level of three Indian villages.

The risk discourse associated with GM technology emerged most prominently following a 1996 study conducted by Dr. Arpad Pusztai at the Rowett Institute in Scotland. This found that rats fed with GM potatoes developed immune system dysfunction and organ abnormalities (Smith 2004: 15–31). There have also been reports of allergic reactions among cultivators, food allergies related to GM soy, sheep deaths and the potential for other, as yet, undetected, unintentional changes within the wider ecological system (Smith 2007). These include concerns for pollen drift, gene drift, 'superweeds' and resistance (Buttel 2005: 313; Weis 2007: 32–33; 73–76).

At a global level, the concern with power relations relates to fears that GM crops are being promoted by multinational corporations, such as Monsanto and Syngenta, whose primary concern is to profit from a corporate takeover of the global food supply (Shiva et al. 1999; Weis 2007: 75). This contributes to anxieties that the commercialisation of agriculture seeks to secure the survival of the powerful, at the expense of the vulnerable, a perspective which raises complex issues of justice, equality and morality within the GM crops debate.

The conflict has resulted in a global rift. A number of GM crops are well established, particularly in the United States, where GM soy, maize and cotton, respectively, accounted for 93, 71 and 90 per cent of the total crop in 2013.[1] GM technology is, however, strongly resisted in other states throughout the world, most notably in Europe (Weis 2007: 77).[2] As Beck (2009: 74–76) highlights, the US government sought the intervention of the World Trade Organisation (WTO) with regard to the European bans, arguing that they represented a barrier to trade and were in violation of free trade agreements. The US government has also sought WTO support in challenging state attempts to regulate and label GM foods (Kuruganti 2006: 4245).

It is noted that the initial context of risk contributes to differences in the way in which GM crops are legitimated worldwide. As Visvanathan and Parmar (2002: 2719) note:

> [i]n Europe and the US, the debate over genetically modified crops has focussed on questions about the environment and food safety. But in developing countries…the possibility that GM crops could make things better…or worse…is a *question of life or death*. [italics in the original]

The governments of a number of countries in the Global South, many of which are among the world's poorest,[3] have chosen to ban GM crops. This is asserted by proponents of the technology as an illustration of how the irresponsible, groundless discourses of risk asserted by an affluent North in relation to GM crops are diffused to an impoverished Global South who would otherwise benefit from them (Prakash and Conko 2006: 44).

This book claims that the controversy surrounding GM crops globally gives rise to legitimation struggles in developing world contexts in relation to their adoption. This relates to the complex 'trade-offs' (Renn 2008: 196) which GM technology involves. Such trade-offs relate not only to assessments by cultivators of whether GM crops alleviate or exacerbate their pre-existing exposure to risk; state governments must also consider trade-offs related to their legitimacy associated with their validation of a controversial technology which is the subject of a risk discourse worldwide.

The assessment of the trade-offs involved in the adoption of GM crops is influenced not only by the initial risk of the context but also by the 'existing distributional practice' (Renn 2008: 133) of key resources and the resulting power relations. It is argued here that the legitimation struggle concerning GM crops relates fundamentally to a challenge to the way in which power is exercised.

Exposure to risk is differentiated in line with access to the key resources of particular contexts and with the associated power relations. For the purposes of the current work, access to resources relates to the means required to negotiate the risks of a particular context. This may include the means of production, but is not restricted to them. Competition to gain greater access to key resources, such as land, water, education and employment represents an attempt to mitigate risk exposure as a concern for the preservation of certain individuals and groups over others. This concern with access to resources highlights the ontological significance of power relations in risk society.

The concern with access to resources is particularly pertinent in high-risk contexts of the Global South given that the benefits and risks of technologies, such as GM crops, are differentiated, along with such access, according to key dimensions of social stratification. Thus, the process of legitimation which the issue of GM crops involves incorporates demands for a 'right to justification' (Forst 2007) of this stratification itself, given that an unequal access to resources influences the way in which the potential risks and benefits of technologies, such as GM crops, are distributed.

The focus on the right to justification has meant that a concern with social justice and equality has become a key focus within the legitimation struggle surrounding GM crops. As Fraser (2008: 16–18) highlights, the idea of justice is multidimensional incorporating not only a concern for redistribution (as access to resources) but also for recognition and political representation. A concern for all three dimensions of justice is present within the struggle to legitimate Bt cotton. This can be seen from the way in which the legitimation struggle concerning GM crops is embedded within a far wider struggle related to the legitimacy of the state and its institutionalisation.

The attempts by cultivators differentiated by land-holding, caste and gender to secure recognition for the right to justification with regard to the existing distribution of resources, and the power relations and differentiated exposure to risk which result from this distribution, form part of the exploration of the legitimation struggle which is undertaken in this book. This is examined through two empirically inseparable, but analytically distinct, elements of legitimation—namely, the legitimation of risk and of democracy. Both of these aspects form the central themes of the book and provide the structural framework with which to analyse the legitimation struggle being undertaken by cultivators with regard to Bt cotton in Telangana, India.

The Legitimation of Risk and New Technology: Bt Cotton in Telangana

Johnson et al. (2006: 55) note that legitimation is a social process which involves a 'collective construction of reality.' The legitimation of risk relates to the way differing constructions of risk are determined and evaluated as a social concern. This involves disputes concerning risk definition (what is the risk, what is its probability of actualisation and its likely impacts?); the trade-offs involved (are certain risks worth taking in order to secure other benefits? Are certain rights worth sacrificing in order to guard against risk?) and the negotiation of risk (how is risk distributed, who is at risk and who is responsible for remedial action when it is actualised?). Answers to these questions must be formulated through discourse at local, national and global levels, discussions which are themselves mediated through power relations at each of these levels.

The manner in which innovations in agriculture are socially diffused and adopted has been of concern to rural sociologists historically. Seminal studies undertaken in the United States included those by Griliches (1957) on the adoption of hybrid corn, Jones (1963) on the diffusion of milking machines and Colman (1968) on the uptake of the mechanical reaper. These studies emphasised the significance of aspects such as profitability and advertising (Griliches), the complexity and comprehensibility of the innovation (Jones), as well as the age, education and socio-economic status of the adopter (Colman). All of these early studies assumed that agricultural innovation was vital to progress and to enhancing the productivity and profitability of agriculture so did not consider risks which may have been associated with the technology itself.

The idea of legitimation with regard to new technologies has more recently emerged within the relatively modern school of thought referred to as Innovation Systems Theory. This developed in the 1980s as part of attempts by economists to theorise socio-technical change (Markard and Truffer 2008: 598). Within this, legitimation is seen as one of the functions involved in seeking to overcome the 'liability of newness' (Geels and Verhees 2011: 911) associated with innovation. Binz et al. (2016: 249) note that, according to the innovation systems approach, 'technology and its institutional context mutually shape each other.'

Technological Innovation System (TIS) theory, which emerged throughout the 1990s from within the general framework of innovation studies, focusses on specific technologies. A technological innovation system is defined as 'a network or networks of agents interacting in a specific technology area under a particular institutional infrastructure to generate, diffuse, and utilise technology' (Carlsson and Stanckiewicz 1991: 94).

Binz et al. (2016: 249) observe that the innovation systems approach has 'treated legitimacy as an outcome of overall system maturation and has not ventured into assessing legitimation as an active process.' The approach, therefore, fails to more deeply analyse the legitimation of technology as a social concern and, as such, largely ignores the impact of power relations on such legitimation.

Again, because the TIS approach deals with the adoption of 'green energy' technologies, such as electric cars or potable water recycling, innovation is regarded as a positive aspect which relies on the public to catch up in terms of acceptance. Given the nature of the technologies which it focusses upon, the TIS approach does not systematically thematise issues concerning the risks of particular technologies or challenges to the validity of controversial technologies. However, as Goncalves (2006: 3) notes, '[a]s technology becomes more contentious, people often question the need for the technology in the first place.' The generally positive outlook on technology in traditional rural sociology and in the TIS approach, therefore, has often failed to address risk in its own terms but has simply regarded the public perception of risk as a stage in the process towards acceptance.

As highlighted, the current study deals with a technology which is strongly contested globally with regard to its risks. While this could be regarded as the anticipated uncertainty associated with an early stage of a new innovation, this study argues that the legitimation process concerning Bt technology is sociologically significant, and the risk discourse cannot be dismissed simply as a form of Neo-Luddism.

This book argues that the legitimation struggle to define the risk of Bt technology represents a challenge to the existing power structure and the way in which knowledge is constructed, as well as the way society itself is constituted, institutionalised and evaluated. It claims that the legitimation struggle explored at the micro-level of the villages has direct implications for the meso-level of the state and the attempts by society to define an institutional structure capable of legitimately legislating for, and regulating, GM technology. This is analysed in Chap. 3.

Hekkert et al. (2007: 429) note that the innovation systems approach 'lacks sufficient attention for the micro level.' The current research asserts that such a focus is important in order to understand the legitimation process as a practical concern. The micro-context of the village explored in this book highlights the way in which the meso-level (state) and macro-level (global) contestation concerning GM technology impacts upon local understandings of the risks of the technology and the normative concern with the way in which power should be exercised through the democratic process.

The analysis explores how these local understandings are mediated through village power relations as part of the legitimation of risk construction and negotiation. The detailed insight into the legitimation process which this book involves is an aspect which rural sociologists and theorists of the TIS approach have tended to neglect given their generally favourable view of technology, and the implication that the perception of risk can be equated simply with resistance to change.

As a site for research into the legitimation of GM crops, India is particularly relevant. Stone (2007: 67) notes:

[i]n the global struggle over genetically modified crops, there are few places where the stakes are more urgent than India, with its 600+ million (mostly small-holder) farmers, its alarming problems in agricultural sustainability, its world-class scientific community, its sophisticated media, and its enormous green NGO sector.[4]

Fieldwork for this book was undertaken between June, 2010 and March, 2011, a duration chosen to coincide with the main agricultural season.[5] The study focusses on Bt cotton, a crop which has been genetically modified to incorporate one or more *Cry* genes from the soil bacterium, *Bacillus thuringiensis* (Stone 2011b: 387). This is purported to render the crop resistant to Lepidopteran bollworms, of which the most destructive is the American bollworm *(Helicoverpa armigera).*[6]

Bt cotton is currently India's only GM crop and was approved for cultivation in 2002. Prior to its official approval, Bt cotton seeds had been illegally adopted by cultivators as early as 1999 (Herring and Rao 2012: 48). Mahyco, a well-established Indian seed company, was granted permission to import 100 grams of Bt cotton seed as part of an agreement with Monsanto as early as 1995 (Scoones 2005: 252). The Review Committee on Genetic Manipulation (RCGM) later approved 40 field trials in nine states (ibid.). It would appear that the wider diffusion of the technology arose as a result of a failure to adequately regulate these field trials (ibid.: 253).

As Stone (2011b: 390) highlights, by 2008, it had become increasingly difficult to obtain non-Bt cotton seed varieties in the Warangal district where the current research is located. This difficulty with access to non-Bt cotton was also noted in the current study and was managed through local NGOs who coordinated demand across a number of villages to ensure that the minimum order quantity for suppliers could be met.

By 2014, 95 per cent of the Indian cotton crop was sown to Bt varieties (Choudhary and Gaur 2015: 4). The adoption rate of Bt cotton in India has made it the fastest-adopted new product in the history of agriculture (Dinham 2001: 7). It has also made Indian farmers record adopters of Bt cotton worldwide.[7] Numerous studies assert the benefits of Bt cotton for Indian cultivators. In research conducted on Bt cotton performance in six Indian states, including the pre-secession state of Andhra Pradesh[8] where the current research is located, Karihaloo and Kumar (2009: 15) reported that the technology had resulted in a 78 per cent increase in crop value due to yield increases and a 14 per cent reduction in pesticide cost.

It was argued that, given the increases in income which the technology permitted, it was of benefit to all classes of cultivator in India, particularly the resource-poor marginal and small cultivators (Choudhary and Gaur 2010: 17). This was also asserted by Subramaniam and Qaim (2010: 304) who argued that 'the main beneficiaries [of Bt cotton were] vulnerable farmers.'

There have also been a range of studies, however, which challenge the increased yields and reduced pesticide use which proponents assert that Bt technology delivers (Qayum and Sakkhari 2005; Kuruganti 2009). Shiva et al. (1999: 601) argue that '[w]hile the benefits of globalisation go to the seeds and chemical corporations through expanding markets, the cost and risks are exclusively borne by the small farmers and landless peasants.'

Shiva et al. (1999: 609) argue that '[c]ontrary to Monsanto's claim, Bt cotton is not 'pest-resistant' but a pesticide producing plant.'[9] These authors claim that the toxicity of the plant is contributing to ecological damage which will exacerbate agrarian risk in the longer-term. This view of the plant's toxicity is supported by ongoing assertions by cultivators and NGOs that Bt cotton is linked to animal deaths (Kuruganti 2006: 4246). Such deaths are dismissed as 'biologically impossible' by the US political scientist, Ronald Herring (2008: 155). Nonetheless, the contestations regarding the risks of the technology have added an epistemic dimension to the legitimation process and heightened concerns related to the vested interests associated with the power struggle which the Bt cotton debate has become. Within this, attempts to construct knowledge with regard to the technology's risks are also subject to legitimation struggle.

The official approval of Bt cotton in India coincided with widespread protests which are ongoing.[10] In 2010, the proposal to extend Bt technology to a food crop—brinjal (aubergine or eggplant), a staple vegetable in India—was greeted with outcry from civil society groups (Gupta 2011: 738–739).[11] Although passed for commercial cultivation by the apex regulatory authority in the country, the Genetic Engineering Approval Committee (the GEAC), a moratorium was later placed on Bt brinjal by the then Minister for the Environment and Forests, Jairam Ramesh. This followed a series of public consultations in six states, including in pre-secession Andhra Pradesh. The multinational, Monsanto, also faces charges of biopiracy given the company's use of local varieties in Bt brinjal's development (Laursen 2012: 11).

The legitimation of Bt cotton crucially hinges upon its ability to alleviate the cultivator distress evidenced by ongoing farmer suicides, particularly in Telangana. Numerous studies have highlighted that these suicides are linked to cultivator indebtedness (Sridhar 2006: 1560; Galab et al. 2009: 169; Deshpande and Arora 2010: 24; Deshpande and Shah 2010: 134; Iyer and Arora 2010: 266; Sreedhar 2010: 227). A source of contestation within the legitimation struggle regarding Bt cotton is, therefore, the extent to which the technology contributes to exacerbating or alleviating farmer suicides and indebtedness.

The purported links between Bt cotton, indebtedness and farmer suicides continue to be debated. Shiva and Jafri (1998) have controversially referred to Bt cotton seeds as 'seeds of suicide.' A study by researchers at Berkeley University (Gutierrez et al. 2015) which analysed suicide data for pre-secession Andhra Pradesh, Maharashtra, Gujarat and Karnataka between 2001 and 2010 found that higher rates of suicide were directly related to Bt cotton adoption in rainfed areas due to the higher costs involved. In a report on this research in *The Hindu* (Venkat 2015), India's former Minister for the Environment and Forests, Jairam Ramesh, claimed '[t]hese findings call for serious discussion relating to [Bt cotton's] long-term sustainability in Indian agriculture.'

Others disagree. In a much referenced paper, Gruère and Sengupta (2011: 316) argue that Bt cotton 'has been very effective overall in India' and is neither 'a necessary nor a sufficient cause of farmer suicides' (ibid.: 317). Gruère and Sengupta (ibid.: 317) also assert that, while Bt cotton 'might have indirectly contributed to farmer indebtedness [via crop failure]…Bt cotton as a technology is not to blame.' Their view is supported by Vasavi (2012: 22) who argues that Bt cotton 'is not the key or singular factor responsible for suicides by agriculturalists.'

Vasavi (2012: 21) notes that suicides result from broader structural problems associated with market liberalisation and the attempts by the state to withdraw from agriculture and agriculturalists. According to Vasavi (ibid.), farmer suicides have been assigned to a 'shadow space' in a 'shining India' (ibid.). An analysis of these structural problems as they pertain to the current research is presented in Chap. 2. It should be recognised, however, that this study seeks to examine the extent to which Bt cotton, as an additional factor, exacerbates or alleviates the already high risks arising from the significant structural problems associated with the contemporary agrarian crisis. Its focus, therefore, is on the impact of Bt cotton on cultivator risk within the context of these already significant structural issues.

The current research is located in the Warangal district of Telangana. Warangal has been described as 'the most controversial district in India' (Herring 2008: 145) with regard to Bt cotton, given that widespread adoption coincides with significant protests. While Telangana, as mentioned, formed part of the wider state of Andhra Pradesh during the research period, it has since divided to form its own state. This book argues that the demands for secession themselves reflected attempts to negotiate the significant risks of the agrarian crisis in the region, risks which are evident from the particularly high numbers of farmer suicides with which Telangana and Warangal are associated.

The fieldwork in Telangana was undertaken in recognition of the view highlighted by Gaurav and Mishra (2012: 7) that few studies of GM crops have 'looked into farmers' risk attitudes.' The methodology also acknowledges Pearson's (2006: 309) perspective that research into Bt cotton in (pre-secession) Andhra Pradesh required an 'extended period of time 'in the field'' using a 'reflexive ethnographic approach' (ibid.). This has been undertaken with the aim of transcending as far as possible the ideological debate between proponents and opponents of the technology. Instead, the study seeks to understand the way in which Bt cotton is both legitimated and delegitimated by cultivators themselves in terms of their own negotiation of risk.

The research explores the experience of 21 cultivators (those with access to their own or leased land) and five landless participants across three Telangana villages during the 2010/2011 cotton season. It examines the way in which risk negotiation is mediated through a village power structure which itself arises from the particular caste composition and land-holding pattern of each village. Gender is included given that, as will be explored in Chap. 2, it is a key dimension along which risk is differentiated given the limited access to land and political influence of females in the Indian context. This power structure determines the way in which Bt cotton is variously legitimated or delegitimated as a strategy of risk negotiation.

Each of the villages has adopted a markedly different approach to the negotiation of risk. Pseudonyms have been used for the villages and participants to seek to preserve their anonymity. This was despite the assurances of participants that actual names could be used. Village pseudonyms have been formulated with the aim of assisting the reader in identifying the cultivation method with which each village is associated as follows:

- **Bantala:** all cotton cultivators have adopted Bt varieties as a result of the emulation of a powerful land-holder in the village. Bantala is associated with a *concentrated power structure* in which access to key resources is concentrated among the powerful.
- **Nandanapuram:** 90 per cent of cotton cultivators adopt Bt varieties. However, 10 per cent have opted for NPM (Non-Pesticide Management), a cultivation practice which prohibits the use of pesticides and Bt cotton varieties.[12] The mixed approach to cultivation reflects the *contested power structure* in the village given the absence of a dominant authority.
- **Orgampalle:** organic cotton is cultivated, and Bt cotton, chemical fertilisers and pesticides are banned as part of the rules of the village. Orgampalle is noted as having a *charismatic power structure* due to the presence of a highly influential village elder.

Given that indebtedness has been found to represent an important factor in the risk of farmer suicides (Vasavi 2012: 23), the study of the legitimation of risk also involves an analysis of the debt levels of participants. This explores the impact of power relations on the legitimation of indebtedness as part of risk negotiation in the villages. It also compares the indebtedness of Bt cotton cultivators with those adopting the alternative methods of organic and Non-Pesticide Management (NPM) practices in order to assess the specific contribution of Bt cotton to the risk exposure of participants within the dimensions of land-holding, caste and gender across the villages.

The Legitimation of Democracy and the Role of the State in Risk Definition

Schmitter and Karl (1991: 4) note that democracy is 'a system of governance in which rulers are held accountable for their actions in the public realm by the citizens, acting indirectly through the competition and cooperation of their elected representatives.' The legitimation of democracy, therefore, involves an evaluation of the extent to which the exercise of power through the democratic process is delivering justice in terms of securing the recognition and representation of the vulnerable. It also entails an assessment of political performance in delivering access to key resources, in order to alleviate exposure to risk, as a collective concern of a given state's citizens.

Both Rousseau ([1762], 1973) and Locke ([1690], 1967: 371) highlight that the function of safeguarding its population from risk is central to a state's legitimacy. As Beetham (2013: 137) questions, 'how can the enormous powers of the state be at all justified, or people be obliged to obey it, unless it fulfil requirements necessary to the society and their own well being, and that it fulfil them effectively?' With regard to the evaluation of this function, Locke ([1690], 1967: 445) claims, 'only the people shall be Judge,' an assertion which is the very basis of the legitimation process itself.

The legitimation of democracy in relation to risk entails attempts to answer the normative question of how exposure to risk should be negotiated, differentiated and addressed as a concern of social justice. The Rawlsian Difference Principle provides a basic criterion by which political legitimacy can be assessed in this regard. This holds that a given social structure should not 'secure attractive prospects for the wealthy unless to do so is to the advantage of those less fortunate' (Rawls [1971], 1999: 65).

The issue of farmer suicides in Telangana, therefore, represents not only a concern for justice but also for the legitimacy of the state itself. As Beck (1994: 18) notes, '[t]echnology can increase productivity but puts legitimacy at risk.' As part of the concern for its legitimacy, the state is required to develop regulatory mechanisms for managing the risks associated with a particular context, as well as welfare mechanisms for securing a redistribution of wealth generated through an unequal access to resources.

The legitimacy of the state is determined through the process of legitimation conducted by its citizens. Within this, the state is itself central in terms of the way in which its intervention in the process is undertaken. This is highlighted by Habermas (1996: 354–359) in his core-periphery model of political will-formation. Habermas (ibid.) argues that decisions taken by the core of the state are legitimate only to the extent that they have been formulated as a result of a discursive process informed by the multiple perspectives on the periphery and can be satisfactorily justified to all of these competing perspectives through the giving of reasons. In this sense, the democratic legitimacy of the state is derived from the way in which the state engages with and supports the process of legitimation which is conducted by its citizens as a concern with self-rule.

With regard to legitimacy, the state must increasingly negotiate the demands of the market. In India, this has been particularly true given the conditionalities imposed as part of structural adjustment for borrowing from the World Bank and International Monetary Fund in 1991. The process of liberalisation arising from this and its impact on state legitimacy is covered extensively in Chap. 3.

The marketisation of agriculture has led to an increasing number of private agents operating in all aspects of agriculture such as the supply of inputs, sale of crops and provision of credit (Vasavi 2012: 82). Attempts at regulation are, however, often challenged by transnational institutions, such as the World Trade Organisation (WTO). This could be seen in the concerns expressed by the WTO regarding India's proposal to label GM foods (Kuruganti 2006: 4245). Such international intervention limits the ability of the state to respond to the demands of its citizens, thereby representing a threat to the state's legitimacy.

Protests against Bt technology were a particular concern in pre-secession Andhra Pradesh, given the state government's already precarious legitimacy. GM crops were highlighted as central to the *Vision 2020* initiative adopted by Chandrababu Naidu's TDP (Telugu Desam Party) government in 1999 as the basis for the state's development. The technology was promoted as part of a wider effort to promote economic growth through the removal of trade barriers and increased diffusion of market principles associated with neoliberal globalisation as the means to alleviating poverty. The latter objective was seen as crucial to addressing the justification claims of the poor in a context of increasing consumerism and wealth.

The analysis highlights, however, that the issue of GM technology which was embedded within this embrace of neoliberal principles became an added threat to the state's legitimacy. This was due to the risk discourse asserted by NGOs and the presence of radical communists who challenged the legitimacy of the state given the ongoing inequality in access to resources over which it presided. The book argues that the conflict concerning Bt cotton shared concerns with the Telangana movement which demanded secession—namely, with regard to the demand for a more equal and just form of development asserted by both sets of protestors.

The issue of Bt cotton, and its implications for state legitimacy, led the pre-secession state to come into conflict with other power-holders in the GM crops debate—most notably, the central government and the multinational, Monsanto. This resulted in the pre-secession government being referred to as 'the most troublesome' in India (Jishnu 2010) with regard to Bt cotton.[13] This was particularly the case in relation to its decision to force Monsanto to reduce the price of Bt cotton seeds which is explored in Chap. 3.

Part of the concern with legitimacy in relation to GM technology is associated with the problematisation of knowledge construction within science itself. Raina (2006: 1624) notes that 'science recognises its weaknesses, its need for social, political and ecological legitimisation over and above economic legitimisation, and the complex nature of decision-making as well as the stakes involved.' The critique of science which the GM crops conflict has given rise to has seen demands for an 'epistemological decentralisation' (Pieterse 2001: 89). This has been increasingly linked to the legitimation of democracy given the perceived need to democratise knowledge construction in response to discourses of risk (Lidskog 2008).

The drive to valorise local knowledge as the means to counteracting the potential reductionism of scientific expertise is, however, countered by the increasing commercialisation of scientific research and its incorporation into efforts to secure economic growth. In India, this marketisation of knowledge has contributed to links between private corporations and the country's publicly funded network of agricultural universities and research institutes (Scoones 2005: 151–153).

This commercialisation has meant that sources of knowledge are increasingly assessed as part of its legitimation. As Shiva (1991: 210) notes, unlike the Green Revolution's publicly-funded development of high-yielding varieties of seeds (HYVs), the research and development of GM crops as part of the Gene Revolution is perceived as driven by private multinationals with a concern for profit rather than philanthropy. Scoones (2005: 35) also notes the growth in privately funded research and development. The reliance of the state on private corporations to drive economic growth through technological innovation made possible by expensive research has led to claims that the state is increasingly acting as an entrepreneurial market agent (Rajan 2006: 101–102).

In India, the ongoing privatisation of knowledge construction was evident from the 2006 Indo-US Knowledge Initiative which saw a more formal alignment between India's public sector agricultural research institutes and US multinationals (Raina 2006). The initiative has brought the legitimacy of the Indian state into question given that the agreement was regarded as being secretly concluded between the United States and Indian governments in ways which circumvented democratic processes and failed to engage stakeholders, such as farmers or local NGOs (Raina 2006: 1622; Vasavi 2012: 151).

The potential for reductionism in the production of knowledge associated with its commodification and privatisation represents a significant threat given humanity's ongoing struggle to negotiate the 'epistemic gap' (Desmond 2014: 13) which risk represents. The fear is that the commercialisation of knowledge will lead to a construction of risk which favours the interests of the powerful and contributes to an ongoing differentiation in risk exposure by marginalising the most vulnerable.

The political turbulence throughout the research period permitted an insight into the inextricable nature of the struggle to legitimate both risk and democracy. Following decades of unrest, which had become particularly intense by the time of the research, it was agreed, in 2014, that Telangana would secede from Andhra Pradesh to become India's twenty-ninth state.

As Kannabiran et al. (2010: 69) note, the secession demands involved chants of 'we want what is ours' amid deliberations over the 'meanings of democracy' (ibid.). These demands had themselves resulted from long centuries of imperial exploitation and an unjust exercise of power which had promoted a highly unequal allocation of resources and differentiated exposure to risk in Telangana (Roosa 2001: 58; Srinivasulu 2002: 6; Haragopal 2010: 52; Maringanti 2010: 35–36).

Demands for secession in Telangana highlighted the way in which the legitimation and delegitmation of Bt cotton in the villages formed part of a wider struggle to secure a just exercise of power and access to resources in the negotiation of risk. This related to attempts to gain recognition and representation for the right to justification for the particular risk exposure of those in Telangana and to secure the resources required to negotiate such risk.

The secession highlighted the view of Beck (1992: 185) that attempts to negotiate risk create the basis for a *new political culture* emerging from 'demands for political participation *outside* the political system' (ibid.) through 'citizens' initiative groups and social movements' (ibid.) [italics in the original]. As the analysis in Chaps. 3 and 4 argues, however, such social movement activity, although emerging from outside the institutionalised political system, is very much a part of the political process. Indeed, as Habermas (1996: 354–359) notes in his core-periphery model, social movements are fundamental to the legitimation struggle which informs the constitution and transformation of the way in which the democratic ideal is translated into political practice.

The analysis in Chap. 7 explores how the institutionalised practice of democracy in the villages is mediated through power relations and linked to patronage opportunities (Corbridge et al. 2013: 176). This serves to reinforce the power structure and to contribute to a depoliticisation of risk, a tendency which was particularly evident in the case of the village of Bantala. Here, participants choose to cultivate Bt cotton, despite their concerns, rather than align with a local NGO in order to secure a supply of non-Bt cotton seeds and adopt an alternative method. This is due to the increased yields and incomes which they argue Bt cotton has procured, as well the potential threat to the village power structure which NGO alignment may represent. This means that the adoption of Bt cotton continues despite the high levels of indebtedness of many Bantala participants.

The presence of NGOs who strongly challenge Bt technology in the villages of Orgampalle (Crops Jangaon) and Nandanapuram (the Deccan Development Society) serves not only to facilitate the adoption of alternative methods through their organisation of the supply of non-Bt seeds and the training they provide on the use of these methods; they also coordinate inter-village mobilisations outside of the institutionalised practice of village democracy in the form of protests against Bt technology. A number of the employees of these NGOs are themselves agricultural scientists and are actively involved in conducting their own fieldwork on Bt cotton.

The linking of democratic practice to the accessing of resources to negotiate risk serves to depoliticise risk. It also contributes to the legitimation of existing power relations and limits the willingness of the vulnerable to demand recognition and representation of their right to justification for their relatively greater exposure to risk. This book argues that the mobilisations organised by NGOs in the villages contribute to enhancing democratic legitimacy in this regard. This is due to the fact that these mobilisations transcend the institutionalised practice of democracy and the power structure which underpins it and seek to challenge both to become more legitimate.

The mobilisations coordinated by NGOs represent an assertion of the right to justification on behalf of the vulnerable in a way which operates beyond the boundary of the village. While the mobilisations against Bt cotton are not entirely inclusive (they exclude, for instance, the landless), they nonetheless extend the demand for recognition of the right to justification to wider sections of the population than a sole reliance on institutionalised democratic practice confined to individual villages, and constrained by the particular power relations of those villages, would permit.

While the issue of the representativeness and accountability of NGOs remains a concern, the book argues that inter-village mobilisations contribute to the representation of those whose access to resources is limited and who, given the impact of local power relations, would otherwise struggle to assert their right to justification and highlight it as a concern for state legitimacy.

Lessons from the Villages in a World at Risk

As a sociological exploration, this book does not provide answers concerning the risks arising from Bt cotton itself in terms of its potential toxicity. The book instead focusses on delivering a socio-economic analysis which examines the legitimation struggle within which cultivators are themselves embedded with regard to Bt technology. This recognises Luhmann's (1993: xi–xii) view that, '[a] sociological investigation cannot seek to take sides, let alone decide the issue. The aim can only be to find out what is going on.'

It is recognised that this is a small-scale study. It does not, therefore, make claims to generalisation. Nonetheless, it is argued that large-scale studies which do make claims on Bt cotton's *overall* effectiveness mask the very 'shadow spaces' where Vasavi (2012) claims the issue of farmer suicides, and the exposure to risk which is causing them, have been consigned. The small scale of this analysis is therefore, it is argued, its primary strength given the more nuanced understanding of the issue which such a micro study permits.

The situation in the village of Nandanapuram, where indebted cultivators adopt Bt cotton even while they protest against it, serves to highlight that the widespread adoption of the technology cannot be assumed to reflect its unproblematic legitimation. While interviews with cultivators suggest that increased yields and reduced pesticide use were associated with Bt cotton for the early adopters in the village, they also indicate that the bases of the technology's initial legitimation have been changing.

These changing bases for the legitimation of Bt cotton relate to assertions of increasing pesticide use and rising cultivation costs, as well as concerns regarding soil erosion, animal deaths and indebtedness. In Bantala, these concerns are traded against the economic benefits in terms of increased yields and incomes which participants assert the technology has enabled in the past (despite the considerable debt levels of many); however, there are indications that this legitimation may be waning in Bantala also given reports of an intended visit by elders in the village to Orgampalle to assess the viability of organic cultivation.

The engagement of Orgampalle and Nandanapuram with NGOs represents an attempt to negotiate risk, to gain wider recognition for the risk exposure of cultivators and to assert their right to justification for this risk. In seeking justification in this way, NGO engagement seeks recognition from the state for questions concerning the justice of the unequal exposure to risk associated with the neoliberal development model.

As Le Mons Walker (2008: 564) highlights, the market-driven approach of neoliberalism in India has seen the number of billionaires rise in a context where welfare expenditures for the poor majority are being reduced.

The resultant inequity in access to resources directly equates to a highly differentiated exposure to risk among India's population which heightens demands for justification and inclusive development.

The study also finds, however, that it is the most powerful in the villages who are spear-heading both Bt cotton's legitimation and delegitimation. Interestingly, it is also the more powerful Bt cotton cultivators in both Nandanapuram and Bantala who are the most indebted. This suggests that the legitimation of Bt cotton coincides with the legitimation of indebtedness and a 'gambling on livelihoods' (Jakimow 2014) as a strategy of risk negotiation. This is due to the fact that, as the analysis highlights, Bt cotton adoption represents a high-cost cultivation method when compared with NPM and organic cultivation methods. While cultivators bear the higher costs (often through loans) hoping that they will result in higher yields and incomes, the erratic climate and variability associated with dryland agriculture in India render this payback far from certain. This high-risk approach is, however, spear-headed by the more powerful, even as the protests against it are.

It has been pointed out that the organisation of protests by the anti-GM NGOs in this study and mobilisations by the more powerful raise ongoing issues of representation; however, the analysis suggests that the discourse of risk asserted in these protests has validity. This is given the significantly higher debt levels of Bt cotton cultivators in all categories of land-holding when compared with their organic and NPM counterparts. There is also the issue of animal deaths which remains unresolved.

While not fully inclusive, such protests serve to pressurise the government, as a concern for its own legitimacy, to deal more adequately with the unequal exposure to risk of cultivators in Telangana than the adoption of Bt technology, and the neoliberal development model through which it is promoted, is permitting. Within this, the book explores how the proposed intention of the new government to grant approval for trials of Bt food crops raises significant concerns for the new state's ongoing legitimacy.[14]

The analysis of the villages gives rise to a number of important insights which are of relevance to the global legitimation struggle in which GM

crops are embedded. Clark (2007: 195) claims, 'legitimacy [is] a constitutive element within a society.' Strydom (2008: 8) also asserts that
attempts to construct and define risk are not simply about 'problem-
solving.' They are also about 'creating and bringing a new world into
being' (ibid.). Thus, the process of legitimation which risk entails is constitutive of local, national and, increasingly, global society.

The research demonstrates the way in which the struggle to legitimate
Bt technology relates not only to efforts to legitimate perspectives on the
risk of the technology itself as an empirical concern but also to the way in
which power is exercised as a normative concern with justification and
justice. This illustrates the view of Pattenden (2005: 1975) that '[t]otalis-
ing views of neoliberal globalisation and its opponents tend to overlook
how processes of 'globalisation' and 'anti-globalisation' are refracted
through specific social, political and material relations.' Such processes of
globalisation give rise to legitimation struggles which 'vary in different
societies' (Abromeit and Stoiber 2007: 35).

The variability of views on legitimacy and the modes of undertaking
protests which challenge such understandings are derived from the different constellations of power, allocation of resources and norms which have
emerged as a result of historical legitimation struggles undertaken within
particular contexts. Despite their particular nature, however, these local
legitimation struggles form part of a wider global legitimation struggle
which is simultaneously testament to, and pivotal to, the constitution of
a global society in which the central concern is the exercise of power in
the negotiation of a world at risk.

The study highlights the significance of NGOs within local legitimation
struggles. It also illustrates, however, the way in which the powerful in the
villages interact differently with NGOs. They either choose to affiliate with
them in order to consolidate their position through a general alleviation of
risk (Orgampalle) or shun them in order to secure the legitimation of their
own positions of power within the village (Bantala). This illustrates that the
way power is itself normatively viewed influences perspectives on how it
should legitimately be exercised. Thus, although challenging the wider
power structure of the state, the decision of the Orgampalle village elder to
affiliate with a local NGO, Crops Jangaon, has served to consolidate the
legitimacy of his position in the village. This is due to the strength of the

legitimacy of the elder's position of power in the village and the general perception that the alignment is safeguarding their welfare against the wider risk of the context in which the village is embedded.

The situation in Orgampalle is relevant to the involvement of NGOs in global deliberations given that the enlistment of NGOs in such debates represents an attempt to secure the legitimacy of the existing power structure, even as NGOs seek to challenge it. It is clear that there are many different constellations of power and that the legitimation of risk involves careful consideration of how these can best be negotiated. It is worth noting, however, that while the involvement of NGOs in global deliberations undoubtedly secures greater recognition of the discourse of risk worldwide, it also contributes to the legitimation of the global power structure through assigning to it the task of defining how such risk should best be addressed.

The Rawlsian Difference Principle which states that a given social structure is unjust when the wealthy benefit in ways which heighten the vulnerability of others highlights the need for a measure of indebtedness to be adopted in conjunction with measures of economic growth at national and global levels. This is vital if the risks being taken by the vulnerable as part of wealth creation are to be properly assessed as a concern for the justice of particular approaches to development and differences in exposure to risk.

The concentrated power structure and strategy of risk negotiation in Bantala, which favoured the powerful, contributed to the village having the highest overall debt exposure in the study. Meanwhile, the negotiation of risk in Orgampalle, where average exposure to the risk of indebtedness was the lowest of the study, suggests the greater effectiveness of a more collective approach to risk negotiation. This was facilitated by the small size and caste homogeneity of the village, as well as the influence of a charismatic village elder whose power was exercised with a view to the security of the collective.

Attempts to apply the lessons from Orgampalle to a global context are recognised as problematic, not least given the significant diversity of world views, values and power relations incorporated in a world at risk. Efforts to approach risk as a collective of humanity must confront the cultural variations arising from differences in power constellations, resource distribution and views on legitimacy and power associated with particular states. Such efforts to approach risk as a global collective must also find ways to negotiate the transnational power relations resulting from, and perpetuating, inequity in the distribution of resources globally.

The research indicates the significant threat which Bt technology and cultivator risk represent to state legitimacy. This is given the questions which risk raises in terms of the legitimacy of the exercise of power by the state itself and the distributional practice over which it presides. Extrapolating this to a global level, it is clear that the materialisation of risk will continue to contribute to significant political upheaval as states attempt to prioritise the risk negotiation of their own citizens as a concern for their legitimacy. This will create difficulties with the formulation of a more collective approach to risk.

The tendency towards a 'boundary problem' (Fraser 2008: 22) of the state with regard to global justice, and attempts by power-differentiated states to negotiate risk individually, is somewhat countered by the emergence of a global legitimation process. Supported by the activity of international NGOs and diffused through travel and media technologies, this is enabling a type of global solidarity which has never before been possible. This global legitimation process contributes to efforts to secure a basic right to justification for the vulnerable which transcends state boundaries. It is also leading to a global concern for knowledge construction in relation to risk due to recognition of the problematic impact of transnational power relations on the way in which knowledge concerning risk is defined, applied and diffused in contexts all over the world.

Beetham (2013: 271) argues that global civil society 'constitutes both an audience for, and an adjudicator of, the legitimacy claims of international institutions.' Local and national legitimation processes which exert pressure on governments for redress of risk are increasingly informed by global concerns in ways which are beginning to foster a more collective (though still imperfect) approach to risk negotiation worldwide. This incorporation of global concerns into national processes of legitimation represents the basis for 'cosmopolitan legitimation' (Beck 2009: 150) in which the citizens of states themselves demand the greater accountability of their governments with regard to globally relevant issues. This forms part of the 'postnational constellation' envisaged by Habermas (2001) and contributes to an emerging global legitimation process in a world at risk.

Chapter Summary

This book is organised as follows:

Chapter 2 looks at the legitimation of risk in the context of village India. This explores the way in which risk exposure and power relations in the Indian context are differentiated primarily along the dimensions of land-holding, caste and gender. The chapter also examines the particular context of risk associated with the Warangal district of Telangana where the villages are located and analyses the legitimation struggle which Bt cotton represents in the region.

Chapter 3 examines the legitimation of democracy. This explores the understandings of democracy in India and the context of pre-secession Andhra Pradesh. The chapter discusses the impact of land-holding, caste and gender on the institutionalised practice of democracy in the villages, most notably with regard to voting in local *panchayat* (council) elections and attendance at the *gram sabha* (village assembly). It argues that the tendency for risk to be depoliticised as a result of the impact of power relations on the institutionalised practice of democracy is mitigated to some extent by NGO mobilisations. The latter represented a particular cause for concern in pre-secession Andhra Pradesh given the wider struggle for legitimacy in which the state was engaged.

Chapter 4 explores the historical struggle for legitimation associated with the Telangana region. The chapter also examines the wider context of risk in which Bt cotton has been adopted in Telangana and Warangal. Finally, it discusses the centrality of concerns related to redistribution, recognition and representation in the demands for the right to justification asserted by the Telangana movement within the struggle for secession. The chapter argues that, as is the case in protests against Bt cotton, concerns for the legitimation of risk and of democracy were inextricably interconnected within these demands for a separate state.

Chapter 5 presents the methodology used to undertake the study. The triangulated, mixed methods approach which the study adopts seeks to gain as broad a perspective as possible on the issue of Bt cotton given that the theme is recognised as controversial and mediated by power relations. The impact of the 'positionality' (Pearson 2006: 308) of the researcher is explored in terms of the power relations involved in the research process itself, as is the way in which reflexivity on this aspect can add new insights. This was particularly the case with regard to working with translators in the current study. The chapter also examines the legitimacy of research into risk, arguing that the very real ethical dilemmas which such research gives rise to need to be carefully balanced with the questionable ethics involved in a failure to undertake such work.

Chapter 6, the first chapter of the analysis, presents a profile of the villages and the power structure with which each is associated, as well as the risk profiles of participants. The analysis seeks to build upon the work developed by Stone (2007: 71) in applying the prestige and conformist biases identified by Henrich (2001: 997) to Bt cotton adoption in Warangal. This explores the impact of power relations on the legitimation and delegitimation of the technology in the villages. Finally, the chapter presents an analysis of debt levels of participants in the villages. This finds that the most powerful Bt cotton cultivators are also the most indebted of the study, a finding which suggests that indebtedness was being legitimated, along with Bt cotton, as part of the neoliberal development model supported by the pre-secession state.

Chapter 7 is the second chapter of the analysis. This explores the legitimation of democracy in the villages in relation to the perspectives of participants on voting and *gram sabha* (village assembly) attendance, as well as their views on state secession. It also looks at the interaction with NGOs in the villages and the links between Bt cotton protests and perceptions of the legitimacy of the state and democratic practice. The chapter argues that the tendency towards a depoliticisation of risk arising from the association of democratic practice with patronage is alleviated somewhat by the inter-village mobilisations coordinated by NGOs. Through the problematisation of power relations and the unequal access to resources asserted as part of the discourse of risk in protests against Bt technology, it is argued that NGOs in the current study contribute to attempts to enhance the legitimacy of democratic practice overall.

Chapter 8 involves a discussion of the key findings. This highlights that Forward Caste medium and semi-medium Bt cotton land-holders are the most indebted in the study. The ostensible wealth creation of these cultivators leads to their cultivation practice being emulated in ways which exacerbate the risk exposure of those with smaller land-holdings and more limited access to resources. The pervasive indebtedness locks cultivators in to Bt cotton adoption, given the ongoing and ever more urgent necessity to secure higher yields in order to clear debts. The chapter argues that the activity of NGOs seeks to assert the right to justification of cultivators made vulnerable through their indebtedness. Finally, the chapter explores how the new state of Telangana will be engaged in a multi-level process of risk negotiation at local, national and global levels in order to secure its own legitimacy. The issue of Bt technology, it is argued, will remain central to this.

Chapter 9 concludes by agreeing with Glover's (2010) assessment that Bt cotton cannot be regarded unproblematically as a 'pro-poor' technology given its erratic results. This view is supported by the high levels of indebtedness with which it is associated in a context where studies have indicated that indebtedness related to crop production is an important factor in farmer suicides (Vasavi 2012: 22–23). While the study acknowledges that Bt cotton is not the sole reason for cultivator borrowing, it argues that it is a contributing factor. It also asserts the need for a measure of indebtedness to be implemented in conjunction with measures of economic growth in order to monitor the risks being taken to create wealth. Finally, the study claims that the legitimation struggle which Bt cotton represents in the villages is a microcosm of the wider legitimation struggle of the global context in which the villages are embedded. This recognition is, it argues, vital for encouraging the type of cosmopolitan knowledge construction required to negotiate risk as a concern of the collective of humanity.

Notes

1. Available at: http://www.gmo-compass.org/eng/agri_biotechnology/ gmo_planting/506.usa_cultivation_gm_plants_2013.html. Accessed on 24/3/2017. Following the US, Brazil and Argentina are the two largest producers of GM maize, soybean and cotton globally. Available at: http://www.genewatch.org/sub-532326. Accessed on 24/3/2017.
2. As of 2015, GM crops were banned in 38 countries worldwide, 28 of which were in Europe. Available at: http://sustainablepulse.com/2015/ 10/22/gm-crops-now-banned-in-36-countries-worldwide-sustainable-pulse-research/#.VqTFr5qLSt9. Accessed on 24/3/2017.
3. These include Algeria, Madagascar, Bhutan, Peru, Belize, Ecuador and Venezuela. Available at: http://sustainablepulse.com/2015/10/22/gm-crops-now-banned-in-36-countries-worldwide-sustainable-pulse-research/#.VqTLJpqLSt_. Accessed on 24/3/2017.
4. As noted by Sainath (2013), the 600 million figure which Stone cites here includes those who are not necessarily farmers but who are dependent on agriculture in an array of other occupations, such as fisheries. Available at: http://www.thehindu.com/opinion/columns/sainath/over-2000-fewer-farmers-every-day/article4674190.ece. Accessed on 24/3/2017.

    According to the 2011 census, there were 118.7 million cultivators and 144.3 million agricultural workers/labourers in India which consists of 31.55 per cent of the total rural population. Sixty per cent of cultivators were small farmers. Available at: http://www.hindustantimes.com/india/how-many-farmers-does-india-really-have/story-431phtct5O9xZS-jEr6HODJ.html. Accessed on 24/3/2017.
5. The start of the *kharif* cropping season varies according to the onset of the monsoon, but the sowing of cotton normally begins in June. Picking of the *kharif* crop starts in late October and continues over a number of months. *Rabi* or winter crops are sown when the monsoon is receding and harvested in spring.

6. Multiple gene hybrids which extend the resistance to further pests, including the Pink and Spotted bollworm, were introduced in 2006.
7. Monsanto (2006) annual report, p. 13. Available at: http://www.monsanto.com/investors/documents/pubs/2006/2006annualreport.pdf. Accessed on 24/3/2017.
8. As highlighted in the Preface, the region of Telangana where the research is located seceded from the wider state of Andhra Pradesh in 2014. Given that the divided state of Andhra Pradesh continues to exist along with the new state of Telangana, this book refers to the 'pre-secession state of Andhra Pradesh' to indicate the undivided state which included the Telangana region.
9. Proponents argue that Bt cotton is toxic only to organisms which contain receptors to Bt proteins—namely, the *Lepidoptera* and *Phthiraptera* classes of insects (Karihaloo and Kumar 2009: 3).
10. Protests against Bt technology have occurred in a number of states, including Karnataka (Scoones 2005), Gujarat (Iyengar and Lalitha 2007) and Maharashtra (Mohanty 2005).
11. Protests against Bt brinjal included a countrywide rally and a day-long fast in 2010. Available at: http://timesofindia.indiatimes.com/india/Countrywide-protest-against-Bt-brinjal/articleshow/5518257.cms Accessed on 24/3/2017.
12. It should be noted that, while Bt cotton is incorporated into the Integrated Pest Management (IPM) approach, it is expressly forbidden as part of NPM practice given the belief that the Bt cotton plant itself incorporates the pesticide and is toxic.
13. Available at: http://business.rediff.com/column/2010/apr/01/guest-bt-cotton-monsanto-is-back-in-courts-over-royalty.htm. Accessed on 25/3/2017.
14. Available at: http://articles.economictimes.indiatimes.com/2015-11-13/news/68252428_1_field-trials-gm-crops-genetic-engineering-appraisal-committee. Accessed on 25/3/2017.

Bibliography

Abromeit, H., & Stoiber, M. (2007). Criteria of democratic legitimacy. In A. Hurrelmann, S. Schneider, & J. Steffek (Eds.), *Legitimacy in an age of global politics* (pp. 35–56). New York: Palgrave Macmillan.
Beck, U. (1992). *Risk society: Towards a new modernity*. London: Sage.

Beck, U. (1994). The reinvention of politics: Towards a theory of reflexive modernization. In U. Beck, A. Giddens, & S. Lash (Eds.), *Reflexive modernization: Politics, tradition and aesthetics in the modern social order* (pp. 1–55). Cambridge: Polity Press.

Beck, U. (1999). *World risk society*. Cambridge: Polity Press.

Beck, U. (2009). *World at risk*. Cambridge: Polity Press.

Beetham, D. (2013). *The legitimation of power*. London: Palgrave Macmillan.

Binz, C., Harris-Lovett, S., Kiparsky, M., Sedlak, D. L., & Truffer, B. (2016). The thorny road to technology legitimation – Institutional work for potable water reuse in California. *Technological Forecasting and Social Change, 103,* 249–263.

Buttel, F. H. (2005). The environmental and post-environmental politics of genetically modified crops and foods. *Environmental Politics, 14*(3), 309–323.

Carlsson, B., & Stanckiewicz, R. (1991). On the nature, function and composition of technological systems. *Journal of Evolutionary Economics, 1,* 93–118.

Choudhary, B., & Gaur, K. (2010). *Bt cotton in India: A country profile*. New York: The International Service for the Acquisition of Agri-biotech Applications (ISAAA).

Choudhary, B., & Gaur, K. (2015). *Biotech cotton in India, 2002 to 2014*. New Delhi: The International Service for the Acquisition of Agri-biotech Applications (ISAAA).

Clark, I. (2007). Legitimacy in international or world society. In A. Hurrelmann, S. Schneider, & J. Steffek (Eds.), *Legitimacy in an age of global politics* (pp. 193–210). New York: Palgrave Macmillan.

Colman, G. P. (1968). Innovation and diffusion in agriculture. *Agricultural History, 42*(3), 173–188.

Corbridge, S., Harriss, J., & Jeffrey, C. (2013). *India today: Economy, politics and society*. Cambridge: Polity Press.

Deshpande, R. S., & Shah, K. (2010). Globalisation, agrarian crisis and farmers' suicides: Illusion and reality. In R. S. Deshpande & S. Arora (Eds.), *Agrarian crisis and farmer suicides* (pp. 118–148). New Delhi: Sage.

Desmond, E. (2014). *The legitimation of risk and democracy: A case study of Bt cotton in Andhra Pradesh, India*. Cork: University College Cork. Available at: https://cora.ucc.ie/handle/10468/1688/

Dinham, B. (2001). GM cotton – Farming by formula? *Biotechnology and Development Monitor, 44,* 7–9.

Forst, R. (2007). *The right to justification*. New York: Columbia University Press.

Fraser, N. (2008). *Scales of justice: Reimagining political space in a globalizing world*. Cambridge: Polity Press.

Galab, S., Revathi, E., & Reddy, P. P. (2009). Farmers' suicides and unfolding agrarian crisis in Andhra Pradesh. In D. N. Reddy & S. Mishra (Eds.), *Agrarian crisis in India* (pp. 164–198). New Delhi: Oxford University Press.

Gaurav, S., & Mishra, S. (2012). *To Bt or not to Bt? Risk and uncertainty considerations in technology assessment.* Mumbai: Indira Gandhi Institute of Development Research.

Geels, F. W., & Verhees, B. (2011). Cultural legitimacy and framing struggles in innovation journeys: A cultural-performative perspective and a case study of Dutch nuclear energy (1945–1986). *Technological Forecasting and Social Change, 78,* 910–930.

Giddens, A. (2003). *Runaway world: How globalization is reshaping our lives.* New York: Routledge.

Glover, D. (2010). Is *Bt* cotton a Pro-Poor Technology? A Review and Critique of the Empirical Record. *Journal of Agrarian Change, 10*(4), 482–509.

Goncalves, M. E. (2006). Risk and the governance of innovation in Europe: An introduction. *Technological Forecasting and Social Change, 73,* 1–12.

Griliches, Z. (1957). Hybrid corn: An exploration in the economics of technological change. *Econometrica, 25*(4), 501–522.

Gruère, G., & Sengupta, D. (2011). Bt cotton and farmer suicides in India: An evidence-based assessment. *Journal of Development Studies, 47*(2), 316–337.

Gupta, A. (2011). An evolving science-society contract in India: The search for legitimacy in anticipatory risk governance. *Food Policy, 36,* 736–741.

Gutierrez, A. P., Ponti, L., Herren, H. R., Baumgartner, J., & Kenmore, P. E. (2015). Deconstructing Indian cotton: Weather, yields and suicides. *Environmental Sciences Europe, 27*(12), 1–17.

Habermas, J. (1996). *Between Facts and Norms: Contributions to a Discourse Theory of Law and Democracy.* Cambridge: Polity Press.

Habermas, J. (2001). *The postnational constellation: Political essays.* Cambridge: Polity Press.

Haragopal, G. (2010). The Telangana people's movement: The unfolding political culture. *Economic and Political Weekly, 45*(42), 51–60.

Hekkert, M. P., Suurs, R. A. A., Negro, S. O., Kuhlmann, S., & Smits, R. E. H. M. (2007). Functions of innovation systems: A new approach for analysing technological change. *Technological Forecasting and Social Change, 74,* 413–432.

Henrich, J. (2001). Cultural transmission and the diffusion of innovations: Adoption dynamics indicate that biased cultural transmission is the predominate force in behavioural change. *American Anthropologist, 103*(4), 992–1013.

Herring, R. J. (2008). Whose numbers count? Probing discrepant evidence on transgenic cotton in the Warangal district of India. *International Journal of Multiple Research Approaches, 2*(2), 145–159.

Herring, R. J., & Rao, N.C. (2012). On the 'failure of Bt cotton': Analyzing a decade of experience. *Economic and Political Weekly, XLVII*(18), 45–53.

Hurrelmann, A., Schneider, S., & Steffek, J. (2007). Introduction: Legitimacy in an age of global politics. In A. Hurrelmann, S. Schneider, & J. Steffek (Eds.), *Legitimacy in an age of global politics* (pp. 1–16). New York: Palgrave Macmillan.

Iyengar, S., & Lalitha, N. (2007). GM cotton in Gujarat: General madness of genuine miracle? *Asian Biotechnology and Development Review, 9*(2), 45–81.

Iyer, K. G., & Arora, S. (2010). Indebtedness and farmers' suicides. In R. S. Deshpande & S. Arora (Eds.), *Agrarian crisis and farmer suicides* (pp. 264–291). New Delhi: Sage.

Jakimow, T. (2014). Gambling on livelihoods: Desire, hope and fear in agrarian Telangana. *Asian Journal of Social Science, 42*, 409–434.

Jangam, C. (2016). Dalit chronicles from the Telugu country. *Economic and Political Weekly, LI*(47), 25–29.

Johnson, C., Dowd, T. J., & Ridgeway, C. L. (2006). Legitimacy as a social process. *Annual Review of Sociology, 32*, 53–78.

Jones, G. E. (1963). The diffusion of agricultural innovations. *Journal of Agricultural Economics, 15*, 387–409.

Kannabiran, K., Ramdas, S. R., Madhusudhan, N., Ashalatha, S., & Kumar, M. P. (2010). On the Telangana trail. *Economic and Political Weekly, XLV*(13), 69–81.

Karihaloo, J. L., & Kumar, P. A. (2009). *Bt cotton in India: A status report* (2nd ed.). New Delhi: Asia-Pacific Consortium on Agricultural Biotechnology and Asia-Pacific Association of Biotechnology Research Institutes.

Knorr-Cetina, K., & Mulkay, M. (Eds.). (1983). *Science observed: Perspectives on the social study of science.* London: Sage.

Kuruganti, K. (2006). Biosafety and beyond: GM crops in India. *Economic and Political Weekly, 41*(40), 4245–4247.

Kuruganti, K. (2009). Bt Cotton and the Myth of Enhanced Yields. *Economic and Political Weekly, XLIV*(22), 29–33.

Latour, B. (1983). Give me a laboratory and I will raise the world. In K. Knorr-Cetina & M. Mulkay (Eds.), *Science observed: Perspectives on the social study of science.* (pp. 141-170). London: Sage.

Laursen, L. (2012). Monsanto to face biopiracy charges in India. *Nature Biotechnology, 30*(1), 11.

Le Mons Walker, K. (2008). Neoliberalism on the ground in rural India: Predatory growth, agrarian crisis, internal colonization, and the intensification of class struggle. *Journal of Peasant Studies, 35*(4), 557–620.

Lidskog, R. (2008). Scientised citizens and democratised science. Re-assessing the expert-lay divide. *Journal of Risk Research, 11*(1–2), 69–86.

Locke, J. ([1690], 1967). *Two treatises of government.* Cambridge: Cambridge University Press.

Luhmann, N. (1993). *Risk: A sociological theory.* New York: Walter de Gruyter.

Maringanti, A. (2010). Telangana: Righting historical wrongs or getting the future right? *Economic & Political Weekly, XLV*(4), 33–38.

Markard, J., & Truffer, B. (2008). Technological innovation systems and the multi-level perspective: Towards an integrated framework. *Science Direct, 37,* 596–615.

Mohanty, B. (2005). 'we are like the living dead': Farmer suicides in Maharashtra, western India. *Journal of Peasant Studies, 32*(2), 243–276.

Pattenden, J. (2005). Trickle-down solidarity, globalisation and dynamics of social transformation in a south Indian village. *Economic and Political Weekly, 40*(19), 1975–1985.

Pearson, M. (2006). 'Science,' representation and resistance: The Bt cotton debate in Andhra Pradesh, India. *The Geographical Journal, 172*(4), 306–317.

Pieterse, J. N. (2001). *Development theory: Deconstructions/reconstructions.* New Delhi: Vistaar.

Prakash, C. S., & Conko, G. (2006). Agricultural biotechnology caught in a war of giants. In J. Entine (Ed.), *Let them eat precaution.* Washington, DC: American Enterprise Institute for Public Policy Research.

Qayum, A., & Sakkhari, K. (2005). *Bt cotton in Andhra Pradesh: A three-year assessment.* Hyderabad: Deccan Development Society.

Raina, R. S. (2006). Indo-US knowledge initiative: Need for public debate. *Economic and Political Weekly, 41*(17), 1622–1624.

Rajan, K. S. (2006). *Biocapital: The constitution of postgenomic life.* London: Duke University Press.

Rawls, J. ([1971], 1999). *A theory of justice: Revised edition.* Cambridge: Harvard University Press.

Reddy, D. N., & Mishra, S. (2009). Agriculture in the reforms regime. In D. N. Reddy & S. Mishra (Eds.), *Agrarian crisis in India* (pp. 3–43). New Delhi: Oxford University Press.

Renn, O. (2008). *Risk governance: Coping with uncertainty in a complex world.* London: Earthscan.

Roosa, J. (2001). Passive revolution meets peasant revolution: Indian nationalism and the Telangana revolt. *Journal of Peasant Studies, 28*(4), 57–94.

Rousseau, J.-J. ([1762], 1973). *The social contract and discourses.* London: J.M. Dent & Sons.

Schmitter, P., & Karl, T. L. (1991). What democracy is…and is not. *Journal of Democracy, 2*(3), Summer 1991, 75-88.

Scoones, I. (2005). *Science, agriculture and the politics of policy: The case of biotechnology in India.* Hyderabad: Orient Longman.

Scoones, I. (2008). Mobilizing against GM crops in India, South Africa and Brazil. *Journal of Agrarian Change, 8*(2 and 3), 315–344.

Shiva, V. (1988). Reductionist science as epistemological violence. In A. Nandy (Ed.), *Science, hegemony and violence* (pp. 232–257). Delhi: Oxford University Press.

Shiva, V. (1991). *The violence of the green revolution: Third world agriculture, ecology and politics.* London: Zed.

Shiva, V., & Jafri, A. (1998). *Seeds of suicide: The ecological and human costs of globalization of agriculture.* New Delhi: Research Foundation for Science, Technology, Ecology.

Shiva, V., Emani, A., & Jafri, A. (1999). Globalisation and threat to seed security: Case of transgenic cotton trials in India. *Economic and Political Weekly, 34*(10–11), 601–613.

Smith, J. M. (2004). *Seeds of deception: Exposing corporate and government lies about the safety of genetically engineered food.* Devon: Green Books.

Smith, J. M. (2007). *Genetic roulette: The documented health risks of genetically engineered foods.* Vermont: Chelsea Green.

Sridhar, V. (2006). Why do farmers commit suicide? The case of Andhra Pradesh. *Economic and Political Weekly, 41*(16), 1559–1565.

Srinivasulu, K. (2002). *Caste, class and social articulation in Andhra Pradesh: Mapping differential regional trajectories.* London: Overseas Development Institute.

Stone, G. D. (2007). Agricultural deskilling and the spread of genetically modified cotton in Warangal. *Current Anthropology, 48*(1), 67–103.

Stone, G. (2011a). Contradictions in the last mile: Suicide, culture, and e-agriculture in rural India. *Science, Technology & Human Values, 36*(6), 759–790.

Stone, G. D. (2011b). Field *versus* farm in Warangal: Bt cotton, higher yields, and larger questions. *World Development, 39*(3), 387–398.

Strydom, P. (2002). *Risk, environment and society.* Buckingham: Open University Press.

Strydom, P. (2008). Risk communication: World creation through collective learning under complex contingent conditions. *Journal of Risk Research, 11*(1–2), 5–22.

Subramaniam, A., & Qaim, M. (2010). The impact of Bt cotton on poor households in rural India. *Journal of Development Studies, 46*(2), 295–311.

Vasavi, A. R. (2012). *Shadow space: Suicides and the predicament of rural India.* New Delhi: Three Essays Collective.

Venkat, V. (2015, June 21). Bt cotton responsible for suicides in rain-fed areas, says study. *The Hindu.*

Visvanathan, S., & Parmar, C. (2002). A biotechnology story: Notes from India. *Economic and Political Weekly, 37*(27), 2714–2724.

Weber, M. ([1968], 1978). *Economy and society: An outline of interpretative sociology* (Vol. 1). London: University of California Press.

Weis, T. (2007). *The global food economy: The battle for the future of farming.* New York: Palgrave Macmillan.

Wynne, B. (2001). Creating public alienation: Expert cultures of risk and ethics on GMOs. *Science as Culture, 10*(4), 445–481.

Web-Sites

Jishnu, L. (2010). Bt cotton: Monsanto is back in courts over royalty, 1/4/2010. http://business.rediff.com/column/2010/apr/01/guest-bt-cotton-monsanto-is-back-in-courts-over-royalty.htm. Accessed 25 Mar 2017.

Sainath, P. (2013). Over 2,000 fewer farmers every day. *The Hindu,* 2/5/2013. http://www.thehindu.com/opinion/columns/sainath/over-2000-fewer-farmers-every-day/article4674190.ece. Accessed 24 Mar 2017.

2

The Legitimation of Risk in the Villages

This chapter explores the legitimation of risk with regard to Bt cotton. It examines the way in which exposure to the wider risks associated with the agrarian crisis in Telangana is differentiated in the villages according to a power structure defined by land-holding, caste and gender. The chapter also introduces the prestige and conformist biases identified by Henrich (2001: 997) and applied by Stone (2007: 71) to Bt cotton adoption in Warangal. These biases are used to examine the way in which the process involved in both the legitimation and delegitimation of Bt cotton is mediated through the village power structure.

The Legitimation of Risk and Village India

The legitimation of risk relates to the way in which decision-making on risk is undertaken in different contexts. As part of this process, the justice of particular constellations of power, and the unequal access to resources which underpins these, as well as the differentiated exposure to risk which this access to resources gives rise to, is assessed. As Beetham (2013: 48)

© The Author(s) 2018
E.L. Desmond, *Legitimation in a World at Risk*,
https://doi.org/10.1007/978-981-10-6065-6_2

notes, 'it is through the possession of, or the privileged access to, property that some people acquire and maintain power over others who lack or are denied such access.'

The application of the concept of legitimation to the area of risk broadens its sociological use beyond its traditional concern with state legitimacy. This is in line with Beetham's (2013: 39) view that the idea of legitimacy cannot 'be limited to the sphere of politics' given that 'it is power itself that morally stands in need of legitimation' (ibid.). Beetham (ibid.) also notes, however, that what is 'political' is, in any case, a focus of interest. This book argues that the decision-making on risk undertaken by Indian cultivators, while not occurring through a formalised political process, is inherently 'political,' not least given the global controversy within which Bt technology is embroiled.

The risk profile of particular contexts will determine which resources are key; the differentiated access to these will serve as the basis, not only for the power structure but also for differentiation in exposure to risk. The unequal access to such resources is, therefore, subject to demands for justification, particularly in high-risk contexts. This is also noted by Renn (2008: 133) who asserts the symbolic nature of risk issues as demands for the 'legitimisation of existing distributional practices…[and] power structures.' As the analysis will highlight, the decision-making of cultivators has significant implications for the legitimacy of local power structures and the differentiated access to resources which defines them.

The legitimation of risk involves justification of additional risk-taking, such as that associated with Bt cotton, within existing distributional practices, as a means of negotiating the risks associated with the distributional practices themselves. This involves careful negotiation of the 'trade-offs' (Renn 2008: 196) associated with the adoption of a technology which is the subject of considerable protests and whose risks are recognised as ambiguous, within an already high-risk context marked by its inequality. Such negotiation entails evaluation of the risks of the technology itself and its likely contribution to alleviating or exacerbating the pre-existing risk of the context. The analysis examines how the decision-making on risk which the legitimation of Bt cotton entails is mediated by power-holders in ways which seek to reinforce their positions of power.

The exploration of the legitimation and delegitimation of Bt cotton in the villages highlights the way in which risks associated with the global 'risk society' identified by Beck (1992, 1999), such as those linked to GM crops, are subject to local legitimation struggles. The process of legitimation in the villages can be seen as a microcosm of the wider struggle to legitimate risk as a global concern. The current study adopts Foucault's (1977: 26) idea of a 'micro-physics of power' to illustrate the way in which power relations impact upon the legitimation of risk as an everyday concern and determine behaviour in local contexts. This recognises Foucault's ([1976], 1994: 35) view that the dynamics of domination and resistance must be seen at the point where power is 'completely invested in its real and effective practices....where it installs itself and produces its real effects.' To this end, the study focusses on the Indian village.

India has been described as 'a land of villages' (Inden 1990: 131). The American political sociologist, Barrington Moore (1966: 318), described the village in pre-colonial India as 'the basic cell of Indian society.' The Indian village has also been the subject of sociological interest for centuries. Karl Marx claimed that the social organisation of villages in ancient India and China represented a distinct 'backward,' pre-capitalist, Asiatic mode of production (Hindess and Hirst 1975: 184).[1] Marx argued that nineteenth-century Asian villages were subject to despotic rulers and ranked one step above barbaric.

According to Marx, the unchanging nature of the village, the absence of class conflict and its resistance to exogenous shocks, as well as the presence of a caste system, rendered Indian society static and unchanging, preventing its progression through capitalism and so on to the advanced stage (in Marx's view) of socialism. Marx ([1867], 2007: 394) claimed, '[t]he structure of the economic elements of [Asiatic] society remains untouched by the storm clouds of the political sky.' Marx believed it was the structure of the village itself which was the source of Asia's purported backwardness. Marx (1853) argued 'these idyllic village communities... restrained the human mind within the smallest possible compass, making it the unresisting tool of superstition, enslaving it beneath traditional rules, depriving it of all grandeur and historical energies.'

In the early colonial reports of the nineteenth century, Indian villages were referred to as 'self-sufficient communities' (Elphinstone 1866: 69–71) or 'village republics.' According to Inden (1990: 132), the latter description was first adopted in a British parliamentary enquiry of 1810 by Sir Charles Metcalfe, a colonial administrator.[2] These depictions supported the idea that Indian villages shared a number of characteristics related to the common ownership of land, the combination of cultivation with village industries and the existence of village councils (*panchayats*) formed of elders who managed issues of justice relevant to the village. Such a view failed to take account of the immense heterogeneity of village arrangements, not only given the scale of the sub-continent but also arising from the significant diversity within regions themselves.

Despite the variability in their composition and internal arrangements, the idea of the self-sufficient village became pivotal to India's quest for independence from British rule, primarily due to the influence of Gandhi.[3] For Gandhi, the village was symbolic of India itself and central to his conceptualisation of the country's future development. In personal correspondence to Nehru, Gandhi (Nehru 1958: 506) argued, 'if India is to attain true freedom…people will have to live in villages, not in towns, in huts, not in palaces.'[4]

Brown (2008: xvii) notes that Gandhi claimed that the small-scale society of the Indian village represented 'a spiritual vision of life which had not been corrupted by materialism.' Gandhi ([1947], 2009: 95) argued, 'the moment you talk [to villagers]…wisdom drops from their lips. Behind the crude exterior, you will find a deep reservoir of spirituality.' As Omvedt (1993: 11) observes, Gandhi's wholesale 'rejection of industrial capitalism led him to formulate…the goal of Indian independence in terms of the regeneration of Indian village society, a small-scale, labour-focussed society.'

The village is also the site where the risks of India's agrarian crisis are negotiated. Sixty-nine per cent of India's population live in rural areas.[5] Cultivators are, however, subject to the significant risks of the Indian agrarian crisis which include water scarcity, the general non-viability of agriculture, ecological degradation, climatic catastrophes and the vagaries associated with the erratic rains of the Indian monsoon (Le Mons Walker 2008: 557; Rao 2009; Deshpande and Shah, 2010; Reddy and Mishra 2009; Singhal 2010).

In his study on risk negotiation in the pre-secession state of Andhra Pradesh, Robert Wade (1994: 1) argued that Indian villages differed 'even within culturally homogenous areas' as a result of their 'ecology, internal social structure, demographic composition, relations with external markets and the apparatus of the state (ibid.).' Wade (ibid.) also claimed that 'variations in scarcity and risk in the vital agricultural sphere explain much of the variation to be found in village organisation.'

The study of the three villages undertaken here highlights the variety between the villages which Wade identifies. Unlike Wade, however, it suggests that it is not differences in the degree of scarcity and risk per se which explain varieties in the village organisation. All of the villages featured in this book share a similar context of risk given that all are involved in negotiating the agrarian risks associated with the district of Warangal in Telangana. Instead, it is argued that the variation in the villages arises primarily from the different constellations of power within the villages themselves. It is this power structure, as determined by the numerical composition of particular caste groupings and their respective access to land, as well as the extent of female mobilisation, which impacts upon the way in which risk is constructed and negotiated, and influences how resources are distributed in response. In this sense, the existing power structure (and the attitudes of power-holders) is a central determinant of how the village organises in response to risk and scarcity and may result in markedly different responses to the same risk.

Forst (2014: ix) notes that the order of justification in a society 'relates to the relative standing of persons to one another.' In terms of the Indian village, the differing power relations which define their structure both arise from, and contribute to, differences in the relative exposure of villagers to risk. Village power relations also contribute to variations in the way these differences in risk exposure are legitimated both as a material and ideological concern. In terms of risk, Beck (1992: 23) describes these differentiated relations as 'social risk positions' where 'some people are more affected than others by the distribution…of risks' (ibid.).

Social risk positions do not simply differentiate exposure to risk; they also differentiate the ability to influence others normatively in terms of the way in which risk should be constructed and negotiated. Thus, as Beetham (2013: 46) highlights, '[a] characteristic attribute of the powerful is their ability to define what the goals of the collective as a whole should be.'

Through influencing the way in which risk is negotiated and constructed as part of its legitimation, the positions of the powerful and the distributional practice which supports their positions are themselves legitimated. This relates to the 'self-fulfilling quality' (Beetham 2013: 61) of the legitimation of power. Evidence of this quality with regard to the power structure is notable in terms of the legitimation of risk. This is given that the unequal distribution of resources underlying the power structure which contributes significantly to differences in the exposure to risk is itself legitimated through the influence of the powerful in risk construction and negotiation.

In the current study, the legitimation of village power structures is directly associated with the selection of cultivation methods as the means to negotiating risk. This does not necessarily result in the legitimation of Bt cotton, however. Instead, both the legitimation and delegitimation of Bt cotton serve to legitimate the power structure of particular villages, even as inter-village protests against the technology serve to problematise the wider power structure of the state within which the villages are embedded. The significance of land-holding, caste and gender, as key dimensions along which power and risk exposure are differentiated in the context of the Indian village, will now be explored.

Land-Holding as a Social Risk Position

The centrality of land to the negotiation of agrarian risk in India is highlighted by Agarwal. Agarwal (2003: 193) argues that there is a well-established relationship between 'the risk of rural poverty and land access.' As Frankel (2005: 97) notes, following India's independence from British colonial rule in 1947, more than one-fifth (22 per cent) of all rural households were landless, a further 25 per cent owned less than one acre and 14 per cent owned uneconomic holdings of between 1 and 2.5 acres. This meant that 61 per cent of rural India was landless or owned marginal holdings, a situation which represented a significant risk to survival in a context where agriculture often represented the sole means of sustenance.

Jawaharlal Nehru, India's first prime minister, attempted to bring about greater equality in land-holding through legislation aimed at land reform. There were significant state-wide variations, however, in the implementation and impact of these reforms. These differences, it is argued here, themselves arose from the particular constellations of power within states. As Kohli (2009: 227) observes, 'certain patterns of leadership, ideology and organisation tend to facilitate [land] distribution, while others do not.' Thus, Kohli (ibid.) notes that, while left-of-centre regimes such as those in Kerala and West Bengal could avoid co-option by propertied groups, other states, such as Uttar Pradesh, dominated by commercial peasant interests, were less successful.

Following independence from the British, zamindars[6] and other intermediaries, such as jagirdars[7] and inamdars,[8] were abolished. As Rudolph and Rudolph (1987: 50) note, the abolition of intermediaries led to the emergence of landlords with various sized holdings who rented land to tenants. Land ceiling legislation, which sought to limit the concentration of land-holding, was introduced in 1972 and 1973, and the practice of bonded labour was abolished under the Bonded Labour System (Abolition) Act of 1976 (Robinson 1988: 29).[9] Despite attempts by landowners to avoid land ceiling legislation through the transfer of holdings and partitioning of land, the policy had some limited success. By the end of December 1976, approximately 1.1 million acres of surplus land had been distributed (Frankel 2005: 551).

Nehru also sought to extend the co-operative approach to agriculture.[10] Nehru (1964: 130) argued that '[b]y forming co-operatives, [farmers can] pool their resources for providing credit and getting their supplies of seeds, implements [and] fertilisers…and can organise the sale of their produce.' Nehru (1964: 131) also envisaged joint cultivation of land where individual cultivators would retain ownership of the land and take their proportional share of what was produced.

Nehru's efforts to broaden the co-operative approach were, however, discredited by existing rural power-holders (Robinson 1988: 270; Varshney 1998: 42; Frankel 2005: 202; Kohli 2009: 86). This was largely through well-publicised charges that Nehru's support for co-operatives represented a plan to collectivise Indian agriculture and threaten the private ownership of land along the lines of the approach adopted by China and Russia (Frankel 2005: 165).

The focus on land reform continued throughout the 1970s and 1980s. An Andhra Pradesh Human Development Report compiled by the Centre for Economic and Social Studies in Hyderabad (CESS 2008: 65) notes that the Constitution of India was amended 13 times for the incorporation of 277 land laws in the Ninth Schedule. These amendments mainly related to the conflict between the fundamental right to private ownership of land enshrined in the Constitution and the directive principle which urged equality in access to resources. This issue was particularly associated with Indira Gandhi's efforts to implement land reforms during her first term in office (1966–77).[11]

Over time, efforts at land reform, as well as factors such as the division of land as a result of inheritance (Walker and Ryan 1990: 184), have contributed to a general decrease in the size of holdings. Reddy and Mishra (2009: 6) note that the average land-holding in India declined from 6.5 acres (2.63 hectares) in 1960–1 to just 2.6 acres (1.06 hectares) by 2003.

As Reddy and Mishra (2010: 47) note, 'small farm size is not itself a constraint on efficiency' in terms of productivity; the disadvantages arise from the reduced ability of marginal and small-holders to access credit, their higher transaction costs (through their reduced ability to obtain bulk discounts) and lower prices for crops (as a result of their reduced bargaining power), as well as limitations in making optimum use of irrigation and new technologies. Galab et al. (2009: 194) argue that small and marginal farmers in dryland regions, such as Telangana, are the most vulnerable because they are expected to bear all the costs and considerable risks of their investments with little to fall back on in the event of crop failure.

It should be recognised that access to land is also associated with access to other assets crucial to the alleviation of agrarian risk. Epstein (1973: 99) noted the heightened exposure to agrarian risk associated with the costs of hiring oxen or tractors for cultivators who lacked ownership of such assets. Wade (1994: 34) observed that tractor ownership was confined to the wealthiest land-owning castes.

Land ownership is also associated with enhanced literacy levels. A report by Dev (2012: 31) noted that literacy rates among agricultural labour in India for 2004–2005 were 34 per cent for the landless, 48 per cent for marginal land-holders and 55 per cent for small-holders. For medium and large land-holders, the rate was 59 per cent (72 per cent for males and 39 per cent for females). Given that education leads to greater opportunities to access off-farm employment and an income to mitigate the risk of crop failure, these figures suggest that the heightened risks associated with restricted land ownership coincide with increased constraints on the ability to escape such risks.

Caste as a Social Risk Position

Bayly (1999: 1) refers to the Indian caste system as an 'elaborately strati-fied social hierarchy that distinguishe[s] India from all other societies.' The term *casta* (derived from the Latin *castus* meaning chaste) was first used in sixteenth-century Spain and Portugal to denote species or breed (Bayly 1999: 105). This book argues that the caste system in India is a highly complex and sophisticated means of legitimating the differenti-ated exposure to risk associated with the power structure which caste represents.[12]

The concept of *varna* relates to the depiction of caste in the vast body of sacred texts referred to as the *Vedas*.[13] The *Veda* texts distinguish between the twice-born *(dwija)* upper castes[14] of Brahmin (priests), Kshatriya (warriors) and Vaishya (traders) and the lower caste, non-twice-born Shudra or servile class. Here, status is equated with ritual purity, and detailed rules of interaction govern relations between upper and lower castes in order to avoid the potential for 'pollution' which lower castes are purported to represent. These include strict rules of con-nubium (referring to marriage) and commensality (related to the sharing of food).[15] So-called Untouchables, and the hill and forest populations, now commonly referred to as 'tribals' or Scheduled Tribes, occupy an ambivalent place below, outside or parallel to this *varna* scheme (Bayly 1999: 9).

These broad *varna* categories are differentiated further through the concept of the *jati*. As Khare (1983: 85) highlights, the *jati* refers to the experience of caste in the 'concrete and factual' domain of everyday social life. The *jati* also formed the basis of the *jajmani* system (Mandelbaum 1970: 160–180), where occupational services were provided in return for a specific proportion of the harvest for the service provider.

As Frankel (2005: 5) observes, the *jati* refers to 'smaller, hereditary, endogamous groups associated with a traditional occupation and related to one another in terms of ritual pollution or purity.' *Jatis* in each region number in the hundreds and can be grouped within the broad *varna* categories (with the exclusion of the *jatis* associated with Untouchability). In the villages, *varna* categories carry little relevance as markers of caste; instead, it is the *jati* or sub-caste which is significant.

The Shudra *varna*, as the non-twice-born service providers, contains the most extensive list of *jatis*.[16] Bayly (1999: 8) notes that the term 'caste' is commonly used to refer to both *varna* and *jati*. As a result of the constitutional recognition of caste, the twice-born upper castes are now referred to as Forward Castes (FCs), the Shudras as Backward Castes (BCs) and former Untouchables as Scheduled Castes (SCs). These categories are in common usage and are particularly associated with the political dimension of caste.

Exposure to the risks of the Indian agrarian context was historically differentiated along caste lines. Brahmins, representing the apex of the caste system, were characterised by their special privilege. As Ramakrishna (1983: 4) notes: '[a]lthough they [Brahmins] were a minority, they commanded unchallenged respect and status and acted as the arbiters of morals in society by virtue of their monopoly over learning and ritualistic supremacy.'

The special privileges which upper castes enjoyed meant that they were in a better position to gain access to land as a result of the allocation of land rights to upper caste *zamindars* and revenue collectors during British rule. Varshney (1998: 45) notes that this meant that landlords came from higher castes while lower castes were often landless.

So-called Untouchables were subject to the most pernicious forms of discrimination. This involved their segregation to the margins of the village.[17] They were also traditionally denied admittance to schools, shops and village shrines and banned from using wells (Freeman 1977: 37). Even the shadow of an individual designated as untouchable was said to be defiling (Bayly 1999: 197). As Bayly (ibid.: 195) highlights, nineteenth-century famine records indicate that those designated as Untouchable were among the first to perish in times of scarcity. Untouchability was abolished under Article 17 of the 1949 Indian Constitution (Epstein 1973: 211). As Epstein (ibid.) notes, however, nobody from a higher caste was ever taken to court for breach of this rule.

Today, the particular constellation of power associated with the caste system is regionally variable and relatively unstable, especially in relation to the shudra category. Fluctuations in access to power and resources associated with particular caste groups have been prompted by legislation (such as the abolition of untouchability), constitutional recognition, land reform, caste mobilisations seeking greater political recognition and representation of particular groups and the reservation of educational and political positions for traditionally deprived castes.[18] This has allowed certain castes to seize opportunities to increase their power locally. As Frankel (2005: 6) notes, 'where nonelite peasant castes [most notably, those associated with the Shudra *varna*] having strength of numbers also owned a large share of the land, they were able to exercise the authoritative role of "dominant landowning castes."'

The analysis highlights that dominant castes, where they exist in villages, exert considerable power over everyday life. The concept of a dominant caste was first developed by the Indian sociologist, M.N. Srinivas. Srinivas (1987: 44) observed that 'members of a dominant caste are in a privileged position vis-à-vis the other local castes, and its leaders wield considerable power. The leaders have the greatest stake in the village and, generally, it is they who organise local activity, whether it be a festival, general protest, or fight. They dominate the traditional council or *panchayat*.'

Weiner (2001: 193–195) notes that, while the ideological basis of caste in Indian society is waning, its significance as a social reality is still very much alive. Indeed, the ongoing significance of caste emerges as a result of the challenges to its legitimacy. Paradoxically, demands for the redress of past injustices associated with caste, and the relevance of such historic injustice to the contemporary struggle of lower castes to gain access to resources to negotiate risk, ensure that caste survives as a primary marker of identity and status in Indian society.

The significant attention which has been given to the uplift of the Scheduled Caste population is evident from the finding that the literacy rate among Scheduled Castes in India rose to 66 per cent in 2011, an 11 per cent increase from 2001.[19] Although traditional caste rankings may have seen considerable fluctuation in terms of the access to power and resources with which they are associated, the analysis highlights that the power relations associated with the overall structure of the caste system remain vitally important as a key dimension of differentiation with regard to risk exposure, construction and negotiation in the contemporary Indian village.

Gender as a Social Risk Position

Omvedt (1993: 12) notes that, historically, the 'self-sufficient village [was] linked to the paternalistic rule of kings.' The legacy of male dominance in the village laid the foundation for the suppression of females, a restraint which was legitimated on religious grounds.[20]

The impact of the ongoing patriarchy in Indian society on the risk exposure of females is evident from the gender divide in literacy rates. According to the 2011 Census data, the literacy rate for males stood at 82 per cent while that for females was just 65 per cent for the country as a whole.[21] The difference was reported to be largely due to the ongoing tendency to have female children married at a young age rather than sending them to school.[22] The generally lower levels of literacy among females limit their ability to mitigate their risk exposure through effective economic and political engagement and contribute to their ongoing marginalisation.

The dependency of females arising from their social exclusion contributes to their social perception as a liability. This both leads to, and is reinforced by, the convention of dowry, a payment made by the wife-givers to the family of the wife-receivers in recognition of the burden which women are perceived as representing (Bayly 1999: 54). Dowry is still very much a feature of Indian life, despite the fact that it was made illegal under the Dowry Prohibition Act of 1961 (Shurmer-Smith 2000: 89).[23] Given the deeply engrained norms associated with gender, however, this legislation has been largely ignored, a situation which highlights the way laws themselves have to be legitimated as culturally valid before they are regarded as binding.

The pressure associated with dowry contributes significantly to the risk exposure of females. Dowry deaths linked to the murder of wives and, increasingly, their suicides as a result of disputed dowry payments, add to the vulnerability of females in Indian society (Tesoriero 2005: 321). The convention of dowry also increases the risk of indebtedness for households with daughters and exacerbates, from the outset, the risk exposure associated with being female. India has one of the world's worst gender gap ratios at 940 females to every 1000 males.[24] This gender gap continues to raise concerns regarding the occurrences of sex-selective abortions and female infanticide (Verma 2005: 28).

The perception of females as adding to the overall risk profile of a household belies their actual contribution to maintaining it. In addition to their unpaid household duties, women are also increasingly engaging in paid labour. Da Corta and Venkateshwarlu (1999: 71) highlight the 'feminisation of agricultural labour.' According to Garikipati (2009: 517), a study conducted in 2001 estimated that 43 per cent of agricultural labourers are women. Despite this, it is noted that women's household status, wages and working conditions remain acutely depressed (ibid.).

Their contribution to agriculture notwithstanding, women's rights to property, most notably land, are poorly defined (Corbridge and Harriss 2000: 209), and land remains subject to patrilineal inheritance (Agarwal 1995: A-39). Agarwal (2003: 184) notes that land reform privileged males, and the issue of land rights for women has remained peripheral. As a result of this restricted access to assets and political power, Krishnaraj (2006: 5376) notes that women are most at risk from food deprivation in times of shortage.

In order to address the marginalisation of women, the 73rd Amendment of the Constitution of India reserved one-third of official government positions for females. This, according to Bose (2010: 10), has 'the potential to fundamentally alter the political landscape and challenge existing power relations.' Buch (2009: 9), however, argues that women's participation in politics is 'still not accepted by large sections of society' and is 'tolerated rather than wanted' (ibid.).

The resistance to women's attempts to address their marginalisation has been associated with an exacerbation of their risk exposure. Since 2007, there has been a rise of 20 per cent in reported rapes in the country (Simon-Kumar 2014: 456). Simon-Kumar (ibid.: 455) argues that the increasing prevalence of violence against women in India is the result of the challenge to patriarchal power structures associated with female assertiveness, as well as their increased upward mobility in a context characterised by a high rate of poverty among males.

Land-holding, caste and gender, therefore, represent key dimensions along which the 'social risk positions' described by Beck (1992: 40) continue to be defined in the Indian context. The legitimacy of these positions is, however, in a state of constant flux due to challenges arising from struggles to gain greater access to resources, along with demands for political recognition and representation for the risk exposure of particular groups. As highlighted, the role of education cannot be underestimated in relation to challenges to social risk positions. The report by Srikrishna et al. (2010: 125) commissioned by the Government of India to investigate the grounds for the secession of Telangana, notes that education 'is one of the main avenues for social mobility in a democratic society and also a means for governments to ensure inclusive growth.'[25]

Sen (1999: 293–294) argues that education permits the mitigation of risk through enhancing economic opportunities. In this way, access to education allows individuals to challenge their initial social risk position through seeking to improve their capacity for risk negotiation within it— for example, through securing off-farm employment to escape rural poverty. Sen (ibid.: 295–296) also notes, however, that education enhances political freedom and develops the ability of individuals, regardless of their status, to instigate social change. This permits the legitimacy of social risk positions themselves to be challenged.

Education also serves as a marker of status in Indian society. This was found in the villages in the current study where having children at a private, fee-paying school added to the standing of particular villagers. The prestige of education was particularly associated with 'English-medium' schools as opposed to those where lessons are conducted through local languages. Thus, Faust and Nagar (2001: 2879) argue that the coveted English-medium education 'helps to create and legitimise the social hierarchies and inequalities associated with the contemporary models of development, even as [it] open[s] doors for individual advancement.'

The use of education as a marker of status, as opposed to the means for challenging the inequity associated with such status differentiation, has led to challenges related to the legitimacy of education itself. This has seen increased scrutiny of the way in which education is being incorporated into pre-existing power struggles and used to reinforce inequalities. The issue of GM crops is both embedded in and has become symbolic of the complex struggle for legitimation associated with ongoing challenges to social risk positions within Indian society and it is to an exploration of the Bt cotton debate which this chapter will now turn.

Bt Cotton and the Legitimation of Risk

Debates concerning the extent of poverty in India, the degree to which levels have improved following the country's much-lauded economic growth, and ongoing problems with measurement, form part of the Great Indian Poverty Debate (Deaton and Kozel 2005). Figures from the Planning Commission (Rangarajan 2014: 5) cite an all-India poverty level of 38 per cent between 2009 and 2010.[26] A report in the *Times of India* (16/7/2010), meanwhile, claimed that 55 per cent of India's citizens (645 million people) were classified as poor. This was according to a new measure, the Multi-dimensional Poverty Index (MPI), which sought to extend the indicators of poverty beyond a concern with income.[27]

Whatever the precise extent of poverty in India, it is clear to a casual visitor to the country that considerable wealth co-exists with significant deprivation. It is also evident that, in such an unequal context, the introduction of a technology which benefits the already wealthy without improving the lot of the poor would directly contravene the Rawlsian Difference Principle which claims that 'inequalities are only just if they result in compensating benefits for everyone' (Rawls [1971], 1999: 13). The Difference Principle here relates to the need for moral justification of technological innovation as a concern for the greater good of society as a whole. As discussed in Chap. 1, this justification requirement represents an attempt to establish the normative validity of a new technology as part of its legitimation.

The need for GM crops in terms of their purported contribution to increasing the yields of food crops came into question during the research period given media reports of food rotting in *godowns* (warehouses), and a Supreme Court order that this should be distributed for free to the poor (*Deccan Chronicle*, 7/9/2010). The response by the then Prime Minister Dr. Manmohan Singh was that the food could not be freely distributed given the impact on prices which would 'destroy the incentive to farmers to produce more' (*Deccan Chronicle*, 7/9/2010). Dr. Singh (ibid.) also argued that the Supreme Court should not get involved 'in the realm of policy formulation.'

The incident highlights that poverty in contemporary India coincides with food surpluses, a fact also noted by Varshney (1998: 7). It also suggests that, contrary to the assertions of GM food proponents, feeding the world's population may have as much to do with distribution, politics and the logic of the economic system as it does with production volumes.

While Bt cotton is not a food crop, its impact on the poor and vulnerable remains central to attempts to secure its normative validity as part of the debate surrounding its legitimation. This relates to assertions by proponents of the potential of Bt cotton to deliver benefits to India's resource-poor cultivators as a result of its purported contribution to increased yields and reduced pesticide costs. The widespread adoption of the technology by India's small and marginal cultivators is often asserted as evidence of its legitimation by the poor themselves. Thus, Shah (2005: 4629) claims that 'farmers...are voting with their feet in favour of the technology.' And the US political scientist, Ronald Herring (2008: 146), observes 'it is hard to imagine farmers spreading a technology that is literally killing them.'

The commercial approval of Bt cotton in 2002 represented the beginning of what was referred to as the Gene Revolution (Shiva 1991: 210; Fukuda-Parr 2007). In a number of ways, the legitimation struggle associated with the Gene Revolution replicates that of the Green Revolution which preceded it given the concern with inequality and injustice which the critique of both involves.

The Green Revolution emerged as a result of the introduction of high-yielding varieties of seeds (HYVs) in the 1960s[28] and the increasingly prevalent need for fertilisers, pesticides and assured irrigation on which these seeds are particularly reliant. The Gene Revolution associated with Bt technology utilises HYV seeds and relies upon the same inputs to which these varieties are responsive.

Brass (1990: 287) highlights that the most notable success of the Green Revolution was in wheat production and, to a lesser extent, rice. Its impact was also highly regionally concentrated—in the south, to (pre-secession) Andhra Pradesh, Tamil Nadu and Kerala and, in the north, to Punjab, Haryana and western Uttar Pradesh (ibid.). The Marxist economist, Utsa Patnaik (2007: 2), argues that the Green Revolution benefitted big farmers and rich peasantry while having 'ruinous effects' on the environment and small-holders. The Green Revolution's contribution to widening economic disparities between cultivators is also asserted by Frankel (1971: 194).

There are marked differences, however, between the two revolutions in terms of the bases for legitimation of both. Shiva (1991: 210) notes that the Green Revolution emerged from India's public sector agricultural universities in conjunction with US foundations with the professed objective of feeding the world (but also in order to negotiate political tensions globally (ibid.)). The Gene Revolution, on the other hand, has been exported from the US private sector and is perceived as driven by the profit motive of globally powerful biotech organisations. The ecological impact of the Gene Revolution is also regarded as more uncertain given that it has moved beyond traditional selective breeding to the manipulation of the genetic structure of the plant itself (in the case of Bt technology, through the insertion of genes from a soil bacterium).

Another significant difference in the legitimation of the Green and Gene Revolutions relates to the changed context in which the process of legitimation occurred/is occurring. Varshney (1998: 48) highlights that the policy change associated with the Green Revolution happened when successive droughts had brought India close to famine, and the country was dependent upon food aid from the United States. The struggle to establish legitimacy for the Gene Revolution, however, is occurring when, as highlighted, food surpluses exist and India is no longer reliant on food aid. Despite the fact that the Gene Revolution is, in some ways, an outgrowth of the Green Revolution, therefore, there are significant differences in the bases of legitimation for both.

It could be argued that the main source of legitimacy for Bt cotton is a direct consequence of the Green Revolution. As Karihaloo and Kumar (2009: 1) note, the bollworm, a major cotton pest, had developed resistance to the pesticides adopted so intensively during the Green Revolution. This had forced farmers to apply as many as 10–16 sprays of pesticides per season (ibid.). The Bt gene was inserted into HYV seeds (Kuruganti 2009: 32), a modification which was purported to render the crop resistant to the bollworm for the first 90 days of the plant's growth, so limiting the need for pesticides.[29]

A number of studies assert that the increased yields and reduced pesticide use associated with Bt cotton cultivation have led to income increases, particularly for marginal and small-holders.[30] Choudhary and Gaur (2010: 20) claim that '[i]n terms of income distribution, all types of households benefit [from Bt cotton], including those below the poverty line.' Subramaniam and Qaim (2010: 295) also argue that Bt cotton leads to increased incomes for poor and vulnerable farmers, as well as for hired female workers, due to the higher yields with which they assert it is associated. In his study in Warangal, Stone (2011: 391) confirms that pesticide use halved between 2003 and 2007. Stone (ibid.) also notes, however, the continuing requirement for pesticides due to a changing pest ecology and the increased incidence of sucking pests.

Further studies have highlighted the technology's variability. This arises not only from India's climatic and geographical diversity but also from its highly differentiated socio-economic context.[31] Glover (2010: 482) argues that the 'performance and impacts of GM crops have…been highly variable, socio-economically differentiated and contingent on a range of agronomic, socio-economic and institutional factors.'

Kuruganti (2009: 31) asserts that parameters such as a good monsoon, low pest pressure and soil type are also highly relevant to cotton yields. Kuruganti (ibid.) argues that, in Gujarat, increases in cotton production are more likely to be linked to the increase in irrigation, and to the HYVs themselves, rather than to Bt technology. Kuruganti (ibid.) further states that 'if pest incidence is low due to climatic and other conditions, there cannot be yield increases due to protection from crop losses [as a result of] insect-resistant varieties.'

The adoption of a new technology, such as Bt cotton, within a rainfed context where the risk of crop loss is high, involves accelerated learning under stress. Traditional seed breeding techniques enabled cultivators to skilfully select and save seeds from crops which they had witnessed growing on their own or another's proximate plot of land over an entire season. This permitted the development of crops uniquely suited to particular conditions. In the case of Bt cotton, however, cultivators must choose from a bewildering variety of Bt cotton seeds—numbered at 880 by 2011 (Kathage and Qaim 2012: 11654)—those which are best suited to the particular soil type and micro-context of their land.

Cultivators must also select the pesticides and fertilisers to which these seeds and the particular pest ecology of the season will be particularly responsive. These choices are often made on the advice of seed dealers who have seen neither the crop nor the plot on which it is being grown. Dealers also have a vested interest in simply promoting the most expensive products. As Vasavi (2012: 92) notes, 'far from the earlier forms of contextualised learning that was environmentally based, recent learning is primarily sourced from agri-business agencies.' Thus, the adoption of the additional risk of Bt cotton involves cultivators in a complex process of decision-making within a high-risk context where the stakes of a failed crop are extremely high.

Given the scale of the risks involved in agriculture in India, there have been attempts to define a crop insurance scheme as the means to alleviating cultivator risk. The sheer complexity of the risks associated with the agrarian context has, however, rendered this difficult (Nair 2010: 20). In a seminal ICRISAT study of risk in six villages in the dryland region of South India, Walker and Ryan (1990: 258) note that 'few farmers demand crop insurance unless voluntary programmes are heavily subsidised.' The authors argue that this is 'probably the best indication that benefits [of crop insurance] as perceived by farmers are small' (ibid.).[32] Agarwal (1980: 100) claimed that 'in the context of a paucity of resources..., crop insurance...is not preferable to the direct utilisation of funds for raising agricultural productivity.'

More recently, attempts at weather-based insurance in India have been fraught with problems of 'moral hazard' (where farmers do not buy insurance until they are sure the weather is adverse) and 'adverse selection' (where the insured has better information than the insurer on the risks involved) (Vyas and Singh 2006: 4592). There are also problems with delays in the settlement of claims given the need to collect and process weather station data across large and highly variable areas (Nair 2010: 19). As such, the scope for weather-based crop insurance in India has remained relatively limited.[33]

This book seeks to explore how the additional risk of Bt cotton is variously legitimated and delegitimated within such a high-risk and complex context. Its approach arises from the recognition that, although vulnerable cultivators are central to the global legitimation struggle regarding

GM crops, their perspectives are rarely directly featured within it. This lacuna has led Scoones (2008: 325) to question 'who [are] these farmers, apparently at the centre of the [GM crops] debate, but too often silent in the discussions?'

The three villages where the research is based are located in the Warangal district of Telangana. Telangana is a semi-arid inland state located on the south-east of peninsular India (see Map 1). The extent of agrarian risk in the state is evident from the high numbers of farmer suicides with which it continues to be associated. Prior to secession, Andhra Pradesh was among the top five states in India for farmer suicides.[34] Within this, however, Telangana is identified by Vasavi (2012: 59) as one of India's 'Suicide Hotspots.'

Both Ramachandran et al. (2010: 8) and Galab et al. (2009: 166–167) cite data produced by the Andhra Pradesh *Rythu Sangam* (APRS), a farmers' association, which claims that Telangana accounted for 66 per cent of the 4403 farmer suicides in the pre-secession state of Andhra Pradesh between 1998 and 2006. Warangal was highlighted as being one of the region's worst affected districts (Rao and Suri 2006: 1546). Galab et al. (2009: 166–167) note that 766 suicides occurred in Warangal between 1998 and 2006. This was significantly higher than the next highest district, Mahbubnagar, in which 467 suicides were recorded.

The profile of farmer suicides in Warangal highlights the significance of land-holding, caste and gender with regard to risk negotiation. A sample survey conducted by Revathi (2009: 217) analysed secondary data on farmer suicides between 2003 and 2004 in four districts of Telangana, including Warangal. This found that the overwhelming majority (91 per cent) of agrarian suicides in Warangal was undertaken by small and marginal farmers. It also found that 97 per cent were male and 67 per cent were from Backward Castes.

Recent government statistics on suicides show that Telangana's heightened risk exposure continues post-secession. In 2014, the new state ranked second in India for farmer suicides (Goyal 2015: 267). Of the 898 farmer suicides reported in the state, 69 per cent were male, and small and marginal cultivators accounted for 55 per cent (ibid.: 283). The state also had the highest rate of female farmer suicides in India at 31 per cent (ibid.: x).

The report also highlighted the high numbers of suicides among cultivators of between 5 and 25 acres.[35] These accounted for 42 per cent of the total farmer suicides in Telangana in 2014 (Goyal 2015: 283). The comparison with Revathi's (2009: 217) analysis of the historical data on suicides in the region in 2003/2004 indicated above suggests that suicides are now becoming more common among larger land-holding categories. This will be examined in more detail in the analysis.

The biases identified by Henrich (2001: 997) in relation to the diffusion of innovations were first applied to the adoption of Bt cotton in Warangal by the US anthropologist, Glenn Stone. These biases are prestige bias 'in which a farmer emulates another on the basis of prestige' (Stone 2007: 71) and conformist bias 'in which a farmer adopts a practice when (and because) it has been adopted by many others' (ibid.). The idea of prestige bias is similar to that of 'Sanskritisation' developed by Srinivas (1966: 6) to explain the way in which caste impacts upon social change in India. Here, 'a 'low' Hindu caste…changes its customs, ritual, ideology and way of life in the direction of a high…caste.'

Stone (2007: 71) argues that these biases give rise to social learning which may contribute to maladaptive cultivation practices and preclude the environmental learning which would allow the risks associated with these practices to become problematised. As highlighted in Chap. 1, Stone (2007: 71) claims that this has contributed to 'agricultural deskilling' as farmers increasingly emulate each other, rather than taking their cues for agriculture from the environment itself.

Stone's analysis of agricultural deskilling in Warangal is in line with a significant body of literature on the relevance of social learning to agricultural innovation. In their study of fertiliser adoption in Ghana, for example, Conley and Udry (2001: 669) note how 'information flows through relatively restricted channels, with each farmer learning about new technology from a few sources.'

Neither Stone nor Conley and Udry, however, explore the micro-level sociological question of which farmers become key sources of learning and why. This is also noted by Kumbamu (2007: 891) who argues that Stone does not take account of the way in which the 'existing agrarian structure' relating to sociological aspects such as 'caste, class, gender, political representation, age and educational level' (ibid.) impact upon agricultural deskilling. The current research seeks to address this.

Aspects of the technology itself, as well as the context in which it is being adopted, are recognised as crucial to the way in which social learning occurs. A study conducted by Munshi (2004: 189) explores the differences between the adoption of HYVs of wheat and rice during the Green Revolution and finds that 'social learning will be restricted in a heterogenous neighbourhood.' Munshi (ibid.: 185) argues that social learning is more likely to occur where there is a perception of fewer 'unobserved individual characteristics' (rice is more sensitive to specific growing conditions than wheat so, according to Munshi, social learning is more likely to occur for wheat). In cases of greater heterogeneity, Munshi (ibid.) therefore claims that individual experimentation, rather than social learning, becomes a more reliable way of negotiating the potential risks associated with innovation.

The situation with Bt cotton is, one of significant heterogeneity—both in terms of the numbers of seed varieties, pesticides and fertilisers and in relation to growing conditions associated with specific plots of land. This complexity must be negotiated by cultivators in conjunction with a virtual absence of state extension services. The gap left by the latter is filled by 'experts' ranging from NGOs advocating a variety of philosophical positions, to seed dealers working on commission, and representatives employed by biotech companies to market Bt technology. In this confusing context, contrary to Munshi, the current analysis highlights that social learning does take place but in a highly circumscribed way—most notably, through the village power structure itself.

It is worth noting that, in Munshi's study, the legitimacy of HYVs themselves was not challenged; there were simply differences in relation to the complexity of the crops to which the innovation was applied. It should also be highlighted that the case of Bt cotton in Warangal does not preclude individual experimentation with different seed varieties, pesticides and fertilisers, and personal observations of the fields of others; however, Bt cotton is a controversial technology whose legitimacy itself is challenged. In this case, the choice of cultivation practice, and the decision on whether or not to legitimate Bt cotton in the villages, relies heavily on social learning which is mediated through the village power structure. Once the decision on an agricultural practice is taken, a degree of individual experimentation and environmental learning occurs, but this is also circumscribed by the power structure of particular villages.

In their study of the adoption of a new sunflower crop by farmers in Mozambique, Bandiera and Rasul (2006: 870) found that social learning relies upon networks defined by mutual trust and reciprocity. Within the context of profound uncertainty which Warangal represents, where cultivation represents a precarious gamble for survival, social networks become, as Renn (2008: 28) highlights, one of the few sources of 'ontological security.' The question then becomes which cultivators are relied upon to become key influencers and why?

This book argues that the legitimation and delegitimation of Bt cotton is inseparable from the legitimation and delegitimation of key influencers as defined by the particular power structure of each village. The inclusion of cultivators who delegitimate Bt cotton highlights a point raised by Kumbamu (2007: 891) who claims that Stone's theory of agricultural deskilling neglects the learning as evidenced in the 'non-adoption and abandonment of Bt cotton' by cultivators in Warangal. This book claims that the legitimation and delegitimation of technological innovation are social processes which are mediated through the village power structure.

This chapter has explored the context of the Indian village as the site for the negotiation of agrarian risk. It has argued that the dimensions of land-holding, caste and gender are key to the differentiation of both power relations and risk exposure. The chapter has further examined the debate concerning Bt cotton and has begun to explore the way in which the risk of suicide in the context of Telangana and Warangal is differentiated according to land-holding, caste and gender. Finally, the chapter has highlighted the significance of the prestige and conformist biases explored by Stone (2007: 71) and argued that the social learning involved in both the legitimation and delegitimation of Bt technology is mediated through village power relations.

Notes

1. According to Marx (Hindess and Hirst 1975: 200), the Asiatic mode of production was characterised by the absence of private land ownership, as well as the self-producing unity of handicrafts and agriculture in the villages.
2. In fact, the idea of the self-sufficiency of the village has been refuted in later literature. Srinivas (1987: 38) argued that villagers would not have been entirely self-sufficient but would have depended upon weekly markets for commodities such as salt, spices, iron, silver and gold.
3. Mohandas Karamchand Gandhi (1869–1948), popularly known as Mahatma (Sanskrit: 'high-souled'), was the high-profile leader of India's freedom movement. Gandhi advocated non violent resistance (*ahimsa*) and *swadeshi* or self-sufficiency as a means of challenging the illegitimate exercise of power associated with colonialism. The idea of *khadi* or homespun cotton was central to this, and the *charkha* or spinning wheel became the emblem of the independence movement.
4. Along with Gandhi, Jawaharlal Nehru (1889–1964) was a key figure in the struggle for India's independence from the British and served as India's first prime minister from 1947 until his death in 1964. He is most associated with his support for Fabian socialism, democracy and science and technology as the basis for India's development.
5. According to 2011 census data, out of a total population of 1.2 billion people in India, 833 million live in rural areas. Available at: http://censusindia.gov.in/2011-prov-results/paper2/data_files/india/paper2_at_a_glance.pdf. Accessed on 25/3/2017.
6. Prior to colonial rule, taxes were collected by *zamindars* on behalf of Mughal imperial rulers.
7. *Jagirdars* were de facto rulers of territory with the right to extract revenue.
8. *Inamdars* received *inam* lands granted as gifts in return for services to the ruler.
9. Bonded labour was defined as a 'system of forced or partly forced labour under which a debtor enters... or is presumed to have entered into an agreement with the creditor' (Robinson 1988: 29). Robinson (1988) describes how the abolition of bonded labour in a village in the Medak district of pre-secession Andhra Pradesh made it more difficult for upper castes to influence the voting behaviour of the lower castes who had previously been in a relation of bonded servitude to them.

10. The co-operative society movement had a long history in pre-independence India. The Co-operative Societies Act, introduced in 1904, sought to tackle rural indebtedness by encouraging cultivators to access cheaper credit through their amalgamation into groups (Madan 2007: 58). Between 1912 and 1929, co-operative societies were extended to areas other than credit provision, such as the sale of produce, purchase of inputs, housing and insurance, and various states passed their own Co-operative Societies Acts. By 1935, there were 84,000 societies (ibid.: 60). The co-operative movement later specialised in sectors, such as dairy and sugar-cane production, often in partnership with states. Co-operative societies also developed among vulnerable groups, such as women, nomads and Scheduled Castes (Baviskar 1987: 565).

11. See Frankel (2005: 442–443 and 464–467) for Indira Gandhi's struggle with the Supreme Court in this regard.

12. This directly challenges the view of the French anthropologist, Louis Dumont. In his classic text, *Homo Hierarchicus*, Dumont (1972) suggests that the central concern of caste with purity and pollution encompassed power and held it in check.

13. The *Vedas* are the earliest records of Indian culture. They are estimated to date to between 2000 and 2500 BC (Nehru, [1946], 2004: 72). The depiction of the power structure of caste as divinely ordained represents an attempt to remove it from temporal legitimation challenges.

14. The upper castes are associated with the wearing of a sacred thread, the *suta*. Only male members of the Brahmin, Kshatriya and Vaishya *varnas* are entitled to undergo the ceremony of *upanayana* where the sacred thread is bestowed upon the bearer (Frankel 2005: 5).

15. A detailed account of these is provided in Dumont's (1972) *Homo Hierarchicus*.

16. According to Srinivas (1987: 38), the essential *jatis* in each village were the carpenter, blacksmith, leather worker, potter, barber and washerman.

17. This segregation continues and is a characteristic feature of Indian villages. Beteille (1971: 26–39), Freeman (1977: 24), Omvedt (1994: 70), Robinson (1988: 83) and Srinivas (2003: 26) all report a similar segregation in villages studied in various states throughout India.

18. The 73rd Amendment of the Constitution of India, ratified in 1992, stipulates that 'a share of *panchayat* (council) seats for Scheduled Castes and Scheduled Tribes in proportion with their respective population shares, would be reserved' (Gibson 2012: 416). As Kohli (2009: 86) claims, the extension of reservations to education places and government jobs has been strongly resisted by upper castes.

19. Available at: http://timesofindia.indiatimes.com/india/SC/STs-take-rapid-strides-close-literacy-gap/articleshow/25536193.cms. Accessed on 25/3/2017.

20. The Sanskrit epic, the *Mahabharata*, a highly influential text in Indian society, states that 'when women are corrupted, confusion of castes arises' (Bayly 1999: 13). This historical and religious legacy has contributed to the legitimation by women of a careful circumscription of the norms associated with their behaviour by males and, notably, by women in relation to each other.

21. Available at http://www.census2011.co.in/literacy.php. Accessed on 25/3/2017.

22. Available at: http://www.census2011.co.in/literacy.php. Accessed on 24/8/2017.

23. The costs associated with dowry have risen significantly in recent years. Shurmer-Smith (2000: 91) notes the emergence of the term 'Maruti marriages' where new Maruti cars are offered in partial settlement of a dowry, particularly in urban areas.

24. Available at: http://www.census2011.co.in/sexratio.php. Accessed on 25/3/2017.

25. The report, commonly referred to as the Srikrishna Report, provides a comprehensive overview of the history of Telangana, the situation with regard to the differential access to resources in the pre-secession state and the demands for justification by those asserting the need for a separate state. It is particularly referred to in Chap. 4 of this book.

26. Chapter 3 of the Planning Commission report provides a useful analysis of the considerable issues with poverty estimation in India. The Planning Commission was dissolved in 2014 by Prime Minister Narendra Modi. It has been replaced by NITI Aayog or the National Institution for Transforming India.

27. The measure was developed by the Oxford Poverty and Human Development Initiative for the United Nations Development Programme (UNDP). It includes indicators such as schooling, child mortality, access to drinking water and assets, as well as income.

28. Unlike seed varieties which pollinate naturally, hybrid or HYV seeds are derived from controlled cross-pollination which selects parent plants for favourable, high-yielding, traits.

29. Bt cotton cultivators are required to plant a 'refuge crop' of non-Bt cotton surrounding the Bt cotton plot to ensure that bollworm feeding on Bt cotton will breed with those feeding on non-Bt cotton and so prevent

resistance from developing (Qayum and Sakkhari 2005: 37). The extent to which this occurred in the two villages which adopted Bt cotton in the current study was variable. This highlights the absence of extension services associated with Bt cotton and the lack of assistance with regard to its, in this case, quite specific cultivation.

30. A selection of papers which support Bt cotton's contribution to poverty alleviation for all classes of land-holder, even in dryland agriculture, include: Srivastava and Kolady (2016), Krishna et al. (2016), Sankaranarayanan and Nalayini (2015), Kathage and Qaim (2012), Gruere and Sengupta (2011), Choudhary and Gaur (2010), Subramanian and Qaim (2010), Karihaloo and Kumar (2009).

31. A selection of studies which highlight Bt cotton's uneven performance and questionable contribution to alleviating the risks of poverty include: Fischer (2016), Dowd-Uribe (2014), Glover (2010), Dev and Rao (2007), Mishra (2007), Morse et al. (2007), Pray and Naseem (2007), Ramanjaneyulu and Kuruganti (2006), Smale et al. (2006).

32. The ICRISAT study was a longitudinal study, initiated in 1975 and concluded in 1985. Two of the villages studied were located in the Mahbubnagar district of Telangana, while the other four were located in the state of Maharashtra.

33. The National Agricultural Insurance Scheme (NAIS) was initiated during the *rabi* crop season of 1999–2000 (Vyas and Singh 2006: 4588). The scheme mainly covers food crops, though some coverage of cotton has been registered (ibid.). Nair (2010: 22) notes that more than 80 per cent of covered farmers are loanees for whom the insurance is compulsory. Vyas and Singh (2006: 4587) claim that, by 2005, 83 per cent of cultivators covered were from Maharashtra, [pre-secession] Andhra Pradesh, Madhya Pradesh, Gujarat, Uttar Pradesh, Orissa and Karnataka. They also estimate that, by 2005, just 3 per cent of cultivators in India were covered under the scheme (ibid.).

34. Farmer suicides began to occur in waves from the 1980s and are particularly associated with India's cotton-producing belt (Deshmukh 2010: 175). The bulk of farmer suicides occurred in pre-secession Andhra Pradesh, Maharashtra, Karnataka, Madhya Pradesh and Chhattisgarh. This became known as India's 'suicide belt' (Le Mons Walker 2008: 572).

35. The report combines the categories of semi-medium (5.1–10 acres) and medium land-holders (10.1–20 acres) used in the current study under the category of medium land-holder (which covers 2–10 hectares or 5–25 acres). It does not, unfortunately, explore the dimension of caste.

Bibliography

Agarwal, A. K. (1980). Agricultural crop insurance in India. In C. Y. Lee (Ed.), *Crop insurance for Asian countries* (pp. 99–113). Bangkok: Food and Agricultural Organisation Regional Office for Asia and the Pacific.

Agarwal, B. (1995). Gender and legal rights in agricultural land in India. *Economic and Political Weekly, XXX*(12), A-39–A-56.

Agarwal, B. (2003). Gender and land rights revisited: Exploring new prospects via the state, family and market. *Journal of Agrarian Change, 3*(1 and 2), 184–224.

Bandiera, O., & Rasul, I. (2006). Social networks and technology adoption in Northern Mozambique. *The Economic Journal, 116*(514), 869–902.

Baviskar, B. S. (1987). Cooperatives and rural development in India. *Current Anthropology, 28*(4), 564–565.

Bayly, S. (1999). *Caste, society and politics in India from the eighteenth century to the modern age.* Cambridge: Cambridge University Press.

Beck, U. (1992). *Risk society: Towards a new modernity.* London: Sage.

Beck, U. (1999). *World risk society.* Cambridge: Polity Press.

Beetham, D. (2013). *The legitimation of power.* London: Palgrave Macmillan.

Beteille, A. (1971). *Caste, class and power: Changing patterns of stratification in a Tanjore village.* London: University of California Press.

Bose, P. (2010). Women's reservation in legislatures: A defence. *Economic and Political Weekly, XLV*(14), 10–12.

Brass, P. R. (1990). *The politics of India since independence.* New York: Cambridge University Press.

Brown, J. M. (2008). *Mahatma Gandhi: The essential writings.* Oxford: Oxford University Press.

Buch, N. (2009). Reservation for women in panchayats: A sop in disguise? *Economic & Political Weekly, XLIV*(40), 8–10.

CESS. (2008). *Human development report 2007: Andhra Pradesh.* Hyderabad: Government of Andhra Pradesh.

Choudhary, B., & Gaur, K. (2010). *Bt cotton in India: A country profile.* New York: The International Service for the Acquisition of Agri-biotech Applications (ISAAA).

Conley, T., & Udry, C. (2001). Social learning through networks: The adoption of new agricultural technologies in Ghana. *American Journal of Agricultural Economics, 83*(3), 668–673.

Corbridge, S., & Harriss, J. (2000). *Reinventing India: Liberalization, Hindu nationalism and popular democracy.* Cambridge: Polity Press.

Da Corta, L., & Venkateshwarlu, D. (1999). Unfree relations and the feminisation of agricultural labour in Andhra Pradesh. *Journal of Peasant Studies, 26*(2), 71–139.

Deaton, A., & Kozel, V. (2005). Data and dogma: The great Indian poverty debate. *The World Bank Research Observer, 20*(2), 177–199.

Deshmukh, N. (2010). Cotton growers: Experience from Vidarbha. In R. S. Deshpande & S. Arora (Eds.), *Agrarian crisis and farmer suicides* (pp. 175–191). New Delhi: Sage.

Dev, S. M. (2012). *Small farmers in India: Challenges and opportunities.* Mumbai: Indira Gandhi Institute of Development Research.

Dev, S. M., & Rao, N. C. (2007). *Socioeconomic impact of Bt cotton.* Hyderabad: Centre for Economic and Social Studies (CESS).

Dowd-Uribe, B. (2014). Engineering yields and inequality? How institutions and agro-ecology shape Bt cotton outcomes in Burkina Faso. *Geoforum, 53,* 161–171.

Elphinstone, M. (1866). *History of India: The Hindu Mohammedan periods.* London: John Murray.

Epstein, S. T. (1973). *South India: Yesterday, today and tomorrow, Mysore villages revisited.* London: Macmillan Press.

Faust, D., & Nagar, R. (2001). Politics of development in postcolonial India: English-medium education and social fracturing. *Economic and Political Weekly, 36*(30), 2878–2883.

Fischer, K. (2016). Why new crop technology is not scale neutral – A critique of the expectations for a crop-based African green revolution. *Research Policy, 45*(6), 1185–1194.

Forst, R. (2014). *Justice, democracy and the right to justification.* London: Bloomsbury.

Foucault, M. ([1976], 1994). Two lectures. In M. Kelly (Ed.), Critique and power: Recasting the Foucault/Habermas debate. London: Massachusetts Institute of Technology, pp. 17–46.

Foucault, M. (1977). *Discipline and punish.* London: Penguin.

Frankel, F. R. (1971). *India's green revolution: Economic gains and political costs.* Princeton: Princeton University Press.

Frankel, F. R. (2005). *India's political economy 1947–2004.* New Delhi: Oxford University Press.

Freeman, J. M. (1977). *Scarcity and opportunity in an Indian village.* California: Cummings Publishing Company.

Fukuda-Parr, S. (Ed.). (2007). *The gene revolution: GM crops and unequal development.* London: Earthscan.

Galab, S., Revathi, E., & Reddy, P. P. (2009). Farmers' suicides and unfolding agrarian crisis in Andhra Pradesh. In D. N. Reddy & S. Mishra (Eds.), *Agrarian crisis in India* (pp. 164–198). New Delhi: Oxford University Press.

Gandhi, M. K. ([1947], 2009). *India of my dreams*. Delhi: Rajpal.

Garikipati, S. (2009). Landless but not assetless: Female agricultural labour on the road to better status, evidence from India. *Journal of Peasant Studies, 36*(3), 517–545.

Gibson, C. (2012). Making redistributive direct democracy matter: Development and women's participation in the gram Sabhas of Kerala, India. *American Sociological Review, 77*(3), 409–434.

Glover, D. (2010). Is Bt cotton a pro-poor technology. *Journal of Agrarian Change, 10*(4), 482–509.

Goyal, L. C. (2015). *Accidental deaths and suicides in India, 2014*. Report from National Crime Records Bureau, New Delhi.

Gruère, G., & Sengupta, D. (2011). Bt cotton and farmer suicides in India: An evidence-based assessment. *Journal of Development Studies, 47*(2), 316–337.

Henrich, J. (2001). Cultural transmission and the diffusion of innovations: Adoption dynamics indicate that biased cultural transmission is the predominate force in behavioural change. *American Anthropologist, 103*(4), 992–1013.

Herring, R. J. (2008). Whose numbers count? Probing discrepant evidence on transgenic cotton in the Warangal district of India. *International Journal of Multiple Research Approaches, 2*(2), 145–159.

Hindess, B., & Hirst, P. Q. (1975). *Pre-capitalist modes of production*. London: Routledge & Kegan Paul.

Inden, R. (1990). *Imagining India*. Oxford: Basil Blackwell.

Karihaloo, J. L., & Kumar, P. A. (2009). *Bt cotton in India: A status report* (2nd ed.). New Delhi: Asia-Pacific Consortium on Agricultural Biotechnology and Asia-Pacific Association of Biotechnology Research Institutes.

Kathage, J., & Qaim, M. (2012). Economic impacts and impact dynamics of Bt (bacillus thuringiensis) cotton in India. *Proceedings of the National Academy of Sciences of the United States of America, 109*(29), 11652–11656.

Khare, R. S. (1983). Normative culture and kinship. In *Essays on Hindu categories, processes and perspectives*. Cambridge: Cambridge University Press.

Kohli, A. (2009). *Democracy and development in India: From socialism to pro-business*. New Delhi: Oxford University Press.

Krishna, V., Qaim, M., & Zilberman, D. (2016). Transgenic crops, production risk and biodiversity. *European Review of Agricultural Economics, 43*(1), 137–164.

Krishnaraj, M. (2006). Food security, agrarian crisis and rural livelihoods: Implications for women. *Economic and Political Weekly, 41*(52), 5376–5388.

Kumbamu, A. (2007). Discussion: Beyond agricultural deskilling and the spread of genetically modified cotton in Warangal. *Current Anthropology, 48*(6), 891–893.

Kuruganti, K. (2009). Bt Cotton and the Myth of Enhanced Yields. *Economic and Political Weekly, XLIV*(22), 29–33.

Le Mons Walker, K. (2008). Neoliberalism on the ground in rural India: Predatory growth, agrarian crisis, internal colonization, and the intensification of class struggle. *Journal of Peasant Studies, 35*(4), 557–620.

Madan, V. (2007). *Co-operative movement in India*. New Delhi: Mittal Publications.

Mandelbaum, D. G. (1970). *Society in India: Volume one continuity and change*. London: University of California Press.

Marx, K. (1853, June 25). The British rule in India. *New York Daily Tribune*.

Marx, K. ([1867], 2007). *Capital* (Vol. 1). New York: Cosimo.

Mishra, S. (2007). *Risks, farmers' suicides and agrarian crisis in India: Is there a way out?* Mumbai: Indira Gandhi Institute of Development Research.

Moore, B., Jr. (1966). *Social origins of dictatorship and democracy: Lord and peasant in the making of the modern world*. Middlesex: Penguin University Books.

Morse, S., Bennett, R., & Ismael, Y. (2007). Inequality and GM crops: A case study of Bt cotton in India. *AgBioforum, 10*(1), 44–50.

Munshi, K. (2004). Social learning in a heterogenous population: Technology diffusion in the Indian green revolution. *Journal of Development Economics, 73*, 185–213.

Nair, R. (2010). Crop insurance in India: Changes and challenges. *Economic and Political Weekly, 45*(6), 19–22.

Nehru, J. (1958). *A bunch of old letters*. New Delhi: Asia Publishing House.

Nehru, J. (1964). *Jawaharlal Nehru's speeches volume 4*. Delhi: Government of India.

Omvedt, G. (1993). *Reinventing revolution: New social movements and the socialist tradition in India*. London: M.E. Sharpe.

Omvedt, G. (1994). *Dalits and the democratic revolution: Dr. Ambedkar and the Dalit movement in colonial India*. London: Sage.

Patnaik, U. (2007). Neoliberalism and rural poverty in India. *Economic and Political Weekly, 42*(30), 3132–3150.

Pray, C. E., & Naseem, A. (2007). Supplying crop biotechnology to the poor: Opportunities and constraints. *Journal of Development Studies, 43*(1), 192–217.

Qayum, A., & Sakkhari, K. (2005). *Bt cotton in Andhra Pradesh: A three-year assessment.* Hyderabad: Deccan Development Society.

Ramakrishna, V. (1983). *Social reform in Andhra (1848–1919).* New Delhi: Vikas Publishing House.

Ramanjaneyulu, G. V., & Kuruganti, K. (2006). Bt cotton in India: Sustainable pest management? *Economic and Political Weekly, XLI*(7), 561–563.

Rangarajan, C. (2014). *Report of the expert group to review the methodology for measurement of poverty.* Delhi: Government of India Planning Commission.

Rao, V. M. (2009). Farmers' distress in a modernizing agriculture – The tragedy of the upwardly mobile: An overview. In D. N. Reddy & S. Mishra (Eds.), *Agrarian crisis in India* (pp. 109–125). New Delhi: Oxford University Press.

Rao, P. N., & Suri, K. C. (2006). Dimensions of agrarian distress in Andhra Pradesh. *Economic and Political Weekly, 41*(16), 1546–1552.

Rawls, J. ([1971], 1999). *A theory of justice: Revised edition.* Cambridge: Harvard University Press.

Renn, O. (2008). *Risk governance: Coping with uncertainty in a complex world.* London: Earthscan.

Revathi, E. (2009). 'Farmers' suicides in Andhra Pradesh: Issues and policy concerns. In S. M. Dev, C. Ravi, & M. Venkatanarayana (Eds.), *Human development in Andhra Pradesh: Experiences, issues and challenges.* Hyderabad: Centre of Economic and Social Studies.

Robinson, M. S. (1988). *Local politics: The law of the fishes: Development through political change in Medak district, Andhra Pradesh (South India).* Delhi: Oxford University Press.

Sankaranarayanan, K., & Nalayini, P. (2015). Performance and behaviour of Bt cotton hybrids under sub-optimal rainfall situation. *Archives of Agronomy and Soil Science, 61*(8), 1179–1197.

Scoones, I. (2008). Mobilizing against GM crops in India, South Africa and Brazil. *Journal of Agrarian Change, 8*(2 and 3), 315–344.

Sen, A. (1999). *Development as freedom.* Oxford: Oxford University Press.

Shah, E. (2005). Local and global elites join hands: Development and diffusion of Bt cotton technology in Gujarat. *Economic and Political Weekly, 40*(43), 4629–4639.

Shiva, V. (1991). *The violence of the green revolution: Third world agriculture, ecology and politics.* London: Zed.

Shurmer-Smith, P. (2000). *India: Globalization and change.* London: Arnold.

Simon-Kumar, R. (2014). Sexual violence in India: The discourses of rape and the discourses of justice. *Indian Journal of Gender Studies, 21*(3), 451–460.

Smale, M., Zambrano, P., & Cartel, M. (2006). Bales and balance: A review of the methods used to assess the economic impact of Bt cotton on farmers in developing economies. *AgBioforum, 9*(3), 195–212.

Srikrishna, B. N., Duggal, V. K., Singh, R., Shariff, A., & Kaur, R. (2010). *Committee for consultations on the situation in Andhra Pradesh.* New Delhi: Government of India.

Srinivas, M. N. (1966). *Social change in modern India.* New Delhi: Orient Longman.

Srinivas, M. N. (1987). *The dominant caste and other essays.* Delhi: Oxford University Press.

Srinivas, M. N. (2003). *Religion and society among the Coorgs of South India.* New Delhi: Oxford University Press.

Srivastava, S. K., & Kolady, D. (2016). Agricultural biotechnology and crop productivity: Macro-level evidences on contribution of Bt cotton in India. *Current Science, 110*(3), 311–319.

Stone, G. D. (2007). Agricultural deskilling and the spread of genetically modified cotton in Warangal. *Current Anthropology, 48*(1), 67–103.

Stone, G. D. (2011). Field *versus* farm in Warangal: Bt cotton, higher yields, and larger questions. *World Development, 39*(3), 387–398.

Subramaniam, A., & Qaim, M. (2010). The impact of Bt cotton on poor households in rural India. *Journal of Development Studies, 46*(2), 295–311.

Tesoriero, F. (2005). Strengthening communities through women's self help groups in South India. *Community Development Journal, 41*(3), 321–333.

Varshney, A. (1998). *Democracy, development and the countryside: Urban-rural struggles in India.* New York: Cambridge University Press.

Vasavi, A. R. (2012). *Shadow space: Suicides and the predicament of rural India.* New Delhi: Three Essays Collective.

Verma, P. (2005). Female infanticide – Not a practice of the past. *Off Our Backs, 35*(3/4), 28–31.

Vyas, V. S., & Singh, S. (2006). Crop insurance in India: Scope for improvement. *Economic and Political Weekly, 41*(43/44), 4585–4594.

Wade, R. (1994). *Village republics: Economic conditions for collective action in south India.* San Francisco: ICS Press.

Walker, T. S., & Ryan, J. G. (1990). *Village and household economies in India's semi-arid tropics.* London: The Johns Hopkins University Press.

Weiner, M. (2001). The struggle for equality: Caste in Indian politics. In A. Kohli (Ed.), *The success of India's democracy* (pp. 193–225). New Delhi: Cambridge University Press.

3

Bt Cotton and the Legitimation of Democracy

This chapter explores how the legitimation of democracy involves attempts to secure justice in the exercise of power with regard to redistribution, recognition and representation in the negotiation of risk. This is depicted using Habermas' (1996: 356) core-periphery model which illustrates the vital function of social movements within the legitimation process which is central to democratic political practice. The chapter examines how the attempt to legitimate Bt cotton is embedded within efforts to enhance democratic legitimacy at local, national and global levels.

The chapter then investigates democratic practice in the villages and explores the way its legitimation is mediated through village power relations. This focusses on two aspects—the electoral process and the *gram sabha* meeting (village assembly). The chapter illustrates how the institutional practice of democracy is associated with attempts to secure patronage in ways which legitimate power structures and depoliticise risk. This means that the challenge to power relations themselves is undertaken through extra-institutional mobilisations which are often coordinated across villages by NGOs. The chapter recognises the significant diversity among NGOs, and the problems of representativeness with which they are associated; however, it argues that, notwithstanding certain concerns regarding their approach to the villages, the NGOs in the current study

© The Author(s) 2018
E.L. Desmond, *Legitimation in a World at Risk*,
https://doi.org/10.1007/978-981-10-6065-6_3

contribute to social movement activity in ways which enhance the legitimacy of democratic practice more generally. This is given their efforts to assert the right to justification of the vulnerable.

The chapter then illustrates how Bt technology was embedded within an embrace of neoliberal globalisation adopted by the pre-secession state of Andhra Pradesh in a bid to alleviate poverty as a concern for its own legitimation. The chapter investigates how this strategy itself presented significant risks to state legitimacy, however; this was given its delegitimation by radical Naxalites and NGOs who asserted the inequality and risks with which they argued the neoliberal approach to poverty reduction was associated.

The Legitimation of Democracy and the Negotiation of Risk

Scanlon (2012: 892) argues that it is through a process of legitimation that we try 'to work out with others ideas that can serve as a common standard of justice in our political lives.' The legitimation of democracy relates to attempts to secure a legitimate exercise of power through the democratic process as a concern for social justice. The degree of legitimacy with which the political system is associated relates to the extent to which it secures the recognition and representation of the vulnerable and facilitates citizens' access to resources as their means to negotiating risk.

The realisation of democratic self-rule and the justification of political power in a country of India's inequality, complexity and diversity has proved notoriously difficult. Kothari (2005: 69) claims that the legitimacy of India's democratic institutions has been slowly eroded since Independence. This 'deinstitutionalisation' (Corbridge and Harriss 2000: 76) is particularly associated with Indira Gandhi's time in office.[1]

Singh and Verney (2003: 16) highlight that mass protests against governments have been a feature of governance in India since Gandhi's time. Singh and Verney (ibid.) note that these rarely spill over from one state to another and '[w]hile there is always the possibility that [the] protests will get out of hand, usually they stop short of violence.' Police violence against nonviolent protestors has, however, been reported.[2]

Corbridge and Harriss (2000: 98) note the increasing 'criminalisation of Indian politics' and the election to office of people with criminal records, most notably since Indira Gandhi's second term in office (1980–1984). In recent decades, Indian politics has also been beset by a series of corruption scandals.[3] Frankel (2005: 726) notes that, throughout the 2000s, India was ranked as one of the most corrupt nations in the world by Transparency International. This situation has led to the Indian state being referred to as the 'Goonda Raj' (Kothari 2005: 71).[4]

And, yet, the world's largest democracy is also noted for its deeply paradoxical nature (Chandhoke and Priyadarshi 2009: ix). The country has a free press, numerous political parties and free elections, with an average voter turnout across the country of around 60 per cent (Sharma 2010: 89). And, as Sharma (ibid.: 67) observes, 'democracy has become such an indelible part of the nation's political consciousness that—despite the disillusionment with 'politics as usual'—most Indians continue to maintain a deep philosophical commitment to [it].' This is particularly true of India's poor and vulnerable who, according to an analysis conducted by Alam (2004: 80), show a higher degree of approval of the idea of democracy than the country's wealthy.

The legitimation of democracy entails assessments of the way in which state power is exercised with regard to the distribution of resources and the recognition and representation of the vulnerable. Such legitimacy derives from the state's perceived capacity to offer its citizens protection from risk as the basis for its existence. This was noted by Rousseau ([1762], 1973: 247) who argued that the 'single will [arising from the social contract between the state and its people was] concerned with the…common preservation and general well-being [of all of its citizens].' Locke ([1690], 1967: 371), too, claimed that the supreme power of the commonwealth should be 'directed to no other *end*, but the *Peace, Safety*, and *public good* of the People' [italics in the original].

Given the core concern for citizen welfare, one of the central tasks of the state, and one of the primary bases for its legitimacy, is the allocation of resources. As Forst (2014b: 24) notes, '[d]emocratic procedures must determine which goods are to be allocated to whom by whom on what scale and for what reasons.' However, the state must also ensure that the process for obtaining or producing the resources required by citizens to negotiate risk does not in itself contribute to their risk exposure. Both the allocation of resources, therefore, and the means taken for their procurement, require justification through a process of legitimation.

The significance of the state to the alleviation of the risk exposure of its citizens was evident in the framing of the Indian constitution. In a speech made to the Constituent Assembly on 22 January 1947, Nehru claimed: '[t]he first task of this Assembly is to free India through a new constitution, to feed the starving people, and to clothe the naked masses, and to give every Indian the fullest opportunity to develop himself according to his capacity.'[5]

The significance of democratic practice to the securing of resources in India has led to its being referred to as a 'patronage democracy' (Corbridge et al. 2013: 176). This association of political practice as the means through which resources can be obtained also no doubt contributes to the observation by the Indian political theorist, Ashis Nandy, that 'the poor seem more committed [to democracy] than the ultra rich.'[6]

As highlighted, the Rawlsian Difference Principle, and its assertion that the wealthy cannot benefit at the expense of the vulnerable, forms the central basis for assessments of the legitimacy of a political system and the particular distribution practice over which it presides. This relates not only to evaluations of the justice of the existing power structure (which incorporates power-holders within the state apparatus itself) and the distribution of resources through which it is defined; it also refers to assessments of the way in which state power is exercised in the treatment of the vulnerable.

Political legitimacy relies upon the ability of the state to justify the power relations and unequal access to resources which result in the differentiated exposure to risk of its citizens. This is of particular concern in India where the risks of poverty are acute, but also highly differentiated, as a result of the inequity of the context and the power relations arising from the dimensions of land-holding, caste and gender discussed in the previous chapter.

Forst (2014a: ix) claims that 'the first question of justice is power.' However, Forst (2007: 194) also argues that individuals owe each other a reciprocal and general right to justification, regardless of their access to resources or the power relations involved.[7] This entails the right to ask for and challenge reasons for why power is being exercised in a particular way. Forst (2014a: 6) claims, therefore, that 'the basic question of justice is not what you have but how you are treated.'

The right to justification, although it may normatively be reciprocally and generally owed irrespective of power relations, is often constrained by a limited access to resources. This impacts upon the ability of vulnerable individuals to assert their right to justification. In risk society, it is this inequality in access to resources which is often the basis for the demand for the right to justification. As will be explored in Chap. 4, the significance of access to resources to assessments of the legitimacy of the exercise of power, as well as the consequences of a failure of justification in this regard, can be seen clearly in the demands for a separate state of Telangana.

Fraser (2008: 16) argues that both redistribution and recognition are central to the achievement of justice. Fraser (ibid.) claims that distributive injustice and misrecognition contribute to marginalisation and vice versa. Such marginalisation, it is argued here, constrains the ability of citizens to demand the right to justification, an aspect which was central to the demands for secession in Telangana. This was particularly significant for cultivators in the region whose right to justification for their risk exposure was not being recognised, as evident from the high numbers of farmer suicides with which Telangana was (and continues to be) associated.

This book argues that the right to justification, which may be reciprocally and generally owed as a normative concern, nonetheless needs to be recognised. Efforts to secure recognition for their right to justification by the vulnerable are often constrained by their limited access to resources which serves to consolidate their marginalisation. And it is here that Fraser's (2008: 17) third dimension of justice, that of representation, is of vital significance. Representation involves demanding recognition of the right to justification on behalf of those whose risk exposure is exacerbated as a result of their limited access to resources. Through such representation, the right to justification can be asserted on behalf of the deprived whose exposure to risk would otherwise fail to become problematised given their lack of resources and limited political power. This will be examined more fully with regard to Telangana's secession in the next chapter.

In terms of risk, marginalisation is also problematic because of the need to secure the perspectives of the vulnerable given the epistemic uncertainty which risk represents. This relates to the potential for the vulnerable to be able to offer insights which could contribute to knowledge with regard to risk or its actualisation. Fricker (2007) argues that the marginalisation of perspectives represents 'epistemic injustice' when people are denied in their *capacity as…knower[s]* (ibid.: 20) [italics in the original]. This form of injustice highlights an epistemic dimension to democratic legitimacy. This relates to the securing of epistemic recognition and representation for the perspectives of the vulnerable with regard to their risk exposure, as well as recognising and representing their demands for justification in relation to the allocation of resources to address it. Thus, with regard to risk, justification entails both a redistributive and an epistemic dimension.

In his core-periphery portrayal of political will-formation, Habermas (1996: 356) claims that the legitimation process can be viewed as a circulation of communicative power between the periphery of civil society and the core of the state. The dynamic interaction between the state and civil society allows any 'consciousness of crisis' (ibid.: 357) to become politicised in ways which render its resolution a direct concern of the state in terms of safeguarding its own legitimacy.

Within the dynamic process of democratic will-formation, the activity of social movements, and the collective mobilisations with which they are associated, is crucial. These movements often involve amalgamations of groups and organisations who come together under a common cause. In this way, as Cohen and Arato (1992: 531) note, social movements help to 'redefine identities, to reinterpret norms, and to develop egalitarian, democratic associational forms.' In risk society, it is through such collective mobilisations that the right to justification of the vulnerable is asserted and democratic legitimacy enhanced. This is also highlighted by Haunss (2007: 161) who claims that 'while challenging the legitimacy of their opponents, social movements may, at the same time, strengthen the legitimacy of the system as a whole.'

Habermas (1996: 370) argues that social movements

> attempt to bring up issues relevant to the entire society, to define new ways of approaching problems, to propose possible solutions, to supply new information, to interpret values differently, to mobilise good reasons and to criticise bad ones.

The mass protests against Bt cotton can be seen as part of the legitimation process which Habermas illustrates. Such protests may involve a few hundred people across a number of villages within a state,[8] or nation-wide protests, such as the 'Monsanto Quit India' day.[9] The differences in scale relate to the government (or core) which is being addressed by the mobilisations—the village protests often appeal to the state government for redress, whereas nationwide protests across a number of states seek recognition from the central government in Delhi and demand that pressure be put on federal governments across the entire state of India to take action.

Global protests against GM crops lack a world government or core but they nonetheless contribute to the pressure exerted on states worldwide.[10] This is noted by Lerche (2008: 241) who highlights the '[a]ttempts to influence national governments through transnational advocacy networks.' As will be discussed, these protests involving states worldwide contribute to an emergent global civil society and global legitimation process. The different scales of the protests illustrate the various levels of the legitimation struggle which the issue of GM crops entails.

Through this political activity which operates outside of the institutionalised political structure yet demands redress from it, it is not simply the legitimacy of Bt technology which is being challenged in terms of the potential risk of the technology itself; the protests also challenge the power structure which supports and promotes the technology (most notably, that associated with the US multinational, Monsanto) and the differentiation in the access to resources and risk exposure through which this power structure is itself defined.

The research in the current study highlights that NGOs are central to the coordination of mass mobilisations against Bt technology in the villages. As Haunss (2007: 160) notes, 'the conceptual differences between social movements...and NGOs are...unclear.' In terms of legitimation, the mobilisations which NGOs coordinate form part of a wider social movement coalition which organise against GM crops and demand social justice. This NGO activity serves to ensure that cultivators in the villages are aware of and have the opportunity to engage in these larger protests. This broader engagement contributes to the problematisation by cultivators of their own risk exposure in ways which ensure that it becomes a direct concern for state legitimacy.

The NGO sector in India is a large and vibrant one. The first official estimate of the number of NGOs in the country was 3.3 million in 2009. This translated into one NGO for every 400 Indians (Shukla 2010).[11]

The significance of NGOs in Indian society is highlighted by Katzenstein et al. (2001: 247) who argue that 'it would be a mistake to ignore the burgeoning NGO sector in any analysis of social activism in India.'

It is recognised that NGOs are highly diverse in their activities, orientation and scale (Adeney and Wyatt 2010: 149). They are also markedly different in their approach to the state. It is noted, too, that the sector has been the subject of much critique in terms of its representation of the vulnerable. This includes the argument by Deshpande (2004: 403) that NGOs serve to depoliticise civil society through their co-opting by the state and their enlistment in rolling out the state's development model.

Mehta (2007: 71) argues that the profusion of NGOs can be seen as part of a trend towards a 'post-democracy' given that they are unaccountable to the people they claim to represent. There are also fears that Indian NGOs in receipt of foreign funding are simply pursuing the interests of foreign donors (Adeney and Wyatt 2010: 149) in ways which themselves threaten the legitimacy of the state. The involvement of NGOs in social advocacy can also mean that issues are re-formulated to come into line with a given NGO's particular agenda in ways which compromise the validity of their representation. In his analysis of advocacy against *Dalit* discrimination in North India, for example, Lerche (2008) highlights how the use of transnational networks by local NGOs led to a redirection of the issue of discrimination into a more generalised one associated with the access of *Dalits* to the employment market. This ignored the underlying structural and cultural problems of the particular context which lay at the root of the core problem of the legitimation of the discrimination itself.

The two NGOs featured in this research, Crops Jangaon (Orgampalle) and the Deccan Development Society (Nandanapuram), are local NGOs with strong links to transnational networks. The latter provides an extensive global research and mobilisation platform which supports the exchange of information on Bt cotton studies and facilitates the coordination of mobilisations worldwide. Notwithstanding a somewhat overbearing approach with regard to the villagers which will be discussed as part of the analysis, the presence of these NGOs enabled cultivators to gain recognition for their very real experience of risk in relation to Bt cotton and with regard to the agrarian crisis more generally. The presence of these NGOs was central to making cultivation options, other than Bt cotton, available and to providingtraining and support in the adoption of these alternative methods. In addition, the mobilisation activities in which these NGOs were engaged across the villages challenged the state's reliance on Bt cotton as the means to poverty reduction through highlighting its risks in conjunction with cultivators. Through this grassroots engagement and nuanced knowledge of local agricultural issues, both NGOs contributed to heightening the awareness of cultivators of their right to justification with regard to their exposure to risk.

The current research highlights that the decision of whether to engage with NGOs must itself be legitimated through the village power structure as part of the concern with risk negotiation and the exercise of power in the villages. Decision-making in this regard defines how the power dynamic associated with NGO involvement is itself legitimated in the villages. The significance of these local power relations with regard to the institutionalised practice of democracy in the villages will now be explored.

Democracy in the Villages

The significance of the village within Indian democratic practice was asserted by Gandhi. In his framing of *gram swaraj* (village self-government), Gandhi (1948) argued that 'true democracy cannot be worked by 20 men sitting at the centre. It has to be worked from below by the people of every village.' Gandhi regarded the village as the primary locus for building awareness of the significance of democratic practice as part of self-rule. Gandhi (1930) claimed: '[m]ere withdrawal of the English is not independence. It [independence] means the consciousness of the average villager that he is the maker of his own destiny, he is his own legislator through his chosen representative.'

The formulation of the three-tier *Panchayat Raj* Institutions (PRIs) at village (Gram Panchayat), block (later Mandal Parishad) and district (Zilla Parishad) levels sought to institutionalise a democratic structure which would not only permit villagers the autonomy to resolve their own issues but also secure their inclusion in the wider decision-making process of the Indian state.[12] The *sarpanch* or head person of the *Gram Panchayat* is elected directly by the electorate in the village, while the President of the Mandal Parishad (intermediate level) and Chairman of the Zilla Parishad (district level) are elected from among the directly

elected members (Jain 2006: 243). The term for official PRI positions is
5 years. The *Gram Panchayat* has the power to levy taxes, while the *sarpanch* is responsible for the maintenance of the village infrastructure and
civic amenities, such as water sources, roads, electrification and drains, as
well as for the organisation of development awareness programmes.[13]

In the terms of this book, the design of the PRI structure aimed at
ensuring the recognition and representation of villagers in highlighting
their risk exposure, and at permitting them to assert their right to justification in defining the way development resources were allocated. As such,
the PRI structure sought to instigate a process of legitimation in the
villages which would secure a fairer distribution of resources and exercise
of power and eventually result in a more just and equitable society.

This hope that the democratic process would bring about a fairer society was asserted by Nehru who believed that the political process as institutionalised in the PRI structure would provide the basis for a 'passive
revolution' (Corbridge and Harriss 2000: 21) in India. Through this
institutionalisation of the democratic process, Nehru hoped that the
inequality of power and access to resources associated with the Indian
context would be challenged through altering the bases for the legitimation of power. In a 1959 letter to his Chief Ministers, Nehru (as cited in
Parthasarathi 1989: 300) argued, '[the PRIs] will lead…to a new set-up
completely in the rural areas. New persons will come to the front.'

This challenge to power structures was envisaged as arising through the
increased demand for justification among India's poor as a result of the
legitimation process associated with democratic political practice. The
PRI structure was also seen as the means for distributing resources in the
form of development programmes to address the significant exposure to
risk of many of India's population and for allowing the state's decision-
making on risk to reflect its recognition of the perspectives of citizens.

The *gram sabha* meeting or village assembly represents the grassroots
level of Indian democracy. It is regarded as central to the Indian democratic process and to Gandhi's vision that villagers would be included in
the political decision-making which impacted upon their lives. It was
envisaged that the assembly would not only secure the access to resources
required to negotiate risk as a material concern; it would also provide the
epistemic, deliberative component of democratic practice which would
allow feedback on risk to be elevated to the relevant state authorities.

The deliberative dimension of village democracy is highlighted by Kumar (2009: 213) who identifies the main functions of the *gram sabha* meeting as being to:

(a) facilitate dialogue;
(b) introduce people to the art of negotiation and collective decision-making;
(c) seek accountability of leaders and demand transparency in the functioning of local-level institutions.[14]

The electoral process was regarded as a further means to challenge village power relations. The potential for the already powerful to dominate the political process was to be averted through a system of reservations aimed at ensuring the political inclusion of the vulnerable. The 73rd Amendment of the Constitution of India, ratified in 1992, stipulates that 'one-third of all *panchayat* seats for women and a share of seats for Scheduled Castes and Scheduled Tribes in proportion with their respective population shares, would be reserved' (Gibson 2012: 416).

It was envisaged that such positive discrimination would provide females and Scheduled Castes and Scheduled Tribes with the means for gaining recognition and access to resources to address the particular risk exposure of their respective groups through the democratic process.[15] The system of reservations included reserved places for the key position of *sarpanch* or elected head of the village council. As Powis (2003: 2618) notes, the position of *sarpanch* is 'the prime locus of village power.'

The institutionalised structure of village democracy was to replace the power structure associated with customary village councils. The latter had been dominated by unelected, most often male, village elders who represented particular caste groupings and managed issues of justice within the village. As Mandelbaum (1970: 327, 329–331) highlights, village elders also represented the village in response to external threats. Pur and Moore (2010: 618) note that, in reality, customary village councils and formal PRI structures tend to co-exist in contemporary village India. Their observation is supported by the current analysis which highlights that many villagers continue to rely on village elders to resolve conflicts and oversee the way in which development funding is being channelled through the PRIs.

The current research indicates the way in which constellations of power vary in each village as a result of the numerical presence of castes and their access to key resources, such as land, irrigation, education and employment. The resulting power relations not only exert significant influence on the way in which particular cultivation methods are variously legitimated and delegitimated; they also determine the practice of democracy itself in terms of how risk exposure in the village is recognised and represented and resources distributed in response to it.

As mentioned in Chap. 2, the power structure in the villages is subject to fluctuation as a result of legitimation struggles. Srinivas (1991: 834) observed that India in the early 1990s was characterised by a caste revolution involving the 'shifting dominance' of castes, as well as a 'feminist revolution' (ibid.: 835). As also noted in Chap. 2, the attempts at land reform, and the caste mobilisations since Independence, have led to the emergence of new dominant castes in the villages.

The current study indicates that dominant castes vary and are determined by the numerical presence of castes in particular villages, as well as their relative land-holding and, to some degree, ritual status. The research illustrates how dominant castes, where they exist, influence risk negotiation in ways which contribute to the legitimation of their power within the villages. This relates to the way in which Bt cotton adoption is variously legitimated and delegitimated in the villages as a means to negotiating risk.

Village power relations also impact upon democratic practice. Rao and Sanyal (2010: 167) note that deliberation in the *gram sabha* is 'a competition between groups and individuals who want a piece of the public pie.' This is also highlighted by Kumar (2006: 209) who highlights the perception that the *gram sabha* 'is the place where poor men and women go seeking some benefits…from various welfare schemes doled out by the state.'[16] The meeting has, therefore, become the primary locus of the 'patronage democracy' described by Corbridge et al. (2013: 176) in the villages.

Attendance at the *gram sabha* is mediated through village power relations. Kumar (2006: 209) found that upper caste females regarded the meeting with disdain, claiming it 'was not the place for them.' And in a study of village democracy in Karnataka, Kulkarni (2012: 160) notes that women were largely absent and that the meeting was 'predominantly a male domain.' Again, in research conducted in Karnataka, Bryld (2001: 158) found that most women did not know what had been decided at the meetings and that their husbands rarely discussed *gram sabha* issues with them.

In Kumar's (2006: 207–208) study, the landless were also found to be generally poorly represented, as were Scheduled Castes, though this varied according to the particular power structures of the villages studied.[17] Kulkarni (2012: 160) notes the way in which '[t]he power of caste and gender, evident in the conduct of the [*gram sabha*] meeting, inhibited democratic participation.' Kulkarni claims that this was evident in the lack of respect paid to the lower caste *sarpanch* and the indifferent treatment of Scheduled Castes during the meeting.

Despite the system of reservations, local power relations also impact upon the electoral process, particularly with regard to the position of the *sarpanch*. Buch (2009: 9) claims that 'women in the reserved seats are there by proxy and that their husbands and male relatives exercise power and responsibility on their behalf.' Powis (2003: 2618) also highlights that *panchayat* elections are increasingly 'contested – unofficially – on party lines.' This means that supporters of the particular political party which the *sarpanch* is seen to represent gain undue access to resources.

The tendency for resources to be allocated in line with local affiliations is also noted by Kumar (2006: 230) who argues that 'benefits [are] distributed only to those who are in the good books of the leaders.' In the case of the vulnerable, these affiliations are highly significant to their ability to negotiate risk. Kulkarni (2012: 155) notes that the accessing of resources from development schemes requires that potential beneficiaries submit a written application or notice requesting such resources. The most vulnerable villagers are often illiterate, however, and rely on the assistance of others, especially the *sarpanch*, to formulate such requests (ibid.: 163). This means that the ability to gain access to resources often necessitates legitimation of the power structure.

The general problem of corruption noted in Indian politics is also evident at village level. Powis (2003: 2620) highlights the well-developed culture of 'money politics' in Indian democratic practice. This was also noted in Robinson's (1988: 174) study in pre-secession Andhra Pradesh where election candidates distributed 'cash and toddy[18] to selected groups in villages in which the vote was expected to be close.' The offering of cash and alcohol in return for votes is also described by participants in the current study.

The research on which this book is based suggests that practices associated with the institutionalisation of democracy in the villages are determined by the way in which power is exercised as a result of the particular composition emerging from the numerical presence of particular castes, as well as their relative land-holding. Villagers in Orgampalle and Nandanapuram also engage in wider mass mobilisations against Bt technology coordinated by NGOs. As highlighted, however, this NGO engagement is itself legitimated through the village power structure.

Through inter-village mobilisations which operate outside of the PRI structures, villagers not only demand justification for the wider power structures beyond the village and challenge the justice of the distribution of patronage and approaches to risk supported by the state; they also problematise the institutionalisation of democratic practice itself. Thus, the legitimation of democracy entails both an institutionalised dimension which focusses on patronage (resources and services) and a non-institutionalised dimension which challenges the structures of politcal practice and the way political power is exercised through these. The particular manner in which these two dimensions are translated into practice and intersect is determined by the operation of power relations in local contexts.

The tendency for institutionalised democratic practice to be monopolised by more powerful villagers, who also have the greatest access to resources, is of particular concern in the case of Bt cotton. The impact of the power structure on risk negotiation in the villages renders it likely that the experience of the more vulnerable participants, and the risks associated with their adoption of Bt technology, can become marginalised in the debate concerning its legitimation.

The marginalisation of the vulnerable with regard to Bt cotton cultivation contributes to the potential that any risks associated with the technology's adoption will become depoliticised. This has implications not only for justice but also for the wider epistemic concern with constructing knowledge related to such risks and taking decisions with regard to the broader application of the technology. In the current study, the activity of NGOs has been key to asserting the right to justification of the vulnerable in this regard. Their involvement seeks to assert the delegitimation of the technology on the grounds not only of its erratic performance and purported contribution to animal deaths; it also asserts the illegitimacy of the power structure and access to resources which differentiate the impact of this variable performance.

Given the inter-village mobilisations with which NGO activity is associated, the issue of Bt technology has had to be carefully negotiated by a state whose legitimacy is assessed both on its ability to provide resources in response to risk and on the perceived justice of the way it exercises its power in protecting its citizens from risk. The issue of Bt technology represented a particular concern for the legitimacy of the pre-secession state of Andhra Pradesh, and this will now be examined.

State Legitimacy and Bt Cotton

Habermas (2008: 330) argues that '[the state] cannot preserve the necessary level of legitimacy in the long run unless a functioning economy fulfills the preconditions for an acceptable pattern of distribution.' This reliance on the economy as part of the concern with legitimacy relates to the state's need to secure the resources required to alleviate the risk exposure of vulnerable groups of its citizens as a concern for its own legitimacy.

Given the significant poverty of much of India, as well as its inequality, the state is obliged to assess complex trade-offs between securing economic growth through supporting globalised trade and justifying the risks and inequality which such a strategy entails. This delicate balancing act of legitimation was particularly significant for the pre-secession Andhra Pradesh government given the scale of farmer suicides in the state.

The occurrence of farmer suicides highlighted the acute vulnerability and risk of cultivators within a state whose legitimacy relied upon its ability to protect its citizens. The suicides suggested the failure of the state to adequately or appropriately recognise and respond to the risk exposure and perspectives of its most vulnerable citizens. They also indicated that the Rawlsian Difference Principle of justice which asserts that the plight of the vulnerable should not be exacerbated through the creation of wealth by the powerful was not being met. Bt technology was embedded within this concern for justice, given that it was supported by the state as the means by which the vulnerable could alleviate their exposure to risk.

As previously highlighted, Bt cotton was officially approved in pre-secession Andhra Pradesh in 2002. The introduction had followed a broader shift to a neoliberal approach to poverty alleviation, spear-headed through a *Vision 2020* initiative launched in 1999 by Chandrababu Naidu's Telugu Desam government. The *Vision 2020* scheme was funded with a major loan, referred to as the Andhra Pradesh Economic Restructuring Project, as part of a 1996 agreement with the World Bank (Frankel 2005: 618).[19] The plan included a commitment to eradicate poverty by 2020. This was to be achieved through the adoption of policies focussed on wealth creation, globalisation and technological innovation (Bandyopadhyay 2001: 900; Frankel 2005: 616–617).

As part of *Vision 2020*, it was envisaged that agriculture's share of employment would be reduced from 70 to 45 per cent by 2020 (Frankel 2005: 623). This would occur through the creation of non-farm employment in a service-led economy (ibid.). Central to the objective of poverty alleviation was the support for the 'twin revolutions of rising expectations and information-communications' (Gupta 2002: 88). The declining emphasis on agriculture was reflected in a reduction in public investment. Galab et al. (2009: 189) note that Gross Fixed Capital Formation (GFCF) in agriculture as a share of total GFCF in the state declined from a peak of 13.83 per cent in 1985–6 to an average of 6–7 per cent in the 1990s.

Biotechnology was emphasised as a key strategy in addressing agrarian risk.[20] The *Vision 2020* document states: 'We will need to be far more aggressive in acquiring and applying advanced technologies in a wide range of fields, including agriculture' (Gupta 2002: 12). The significance of biotechnology to the pre-secession state's development plans was evident from the construction of a 600 square kilometre 'Genome Valley' (Rajan 2006: 77) in and around Hyderabad.[21] The development, which claimed to be the 'biotech hub' of India (ibid.), aimed to attract multinationals involved in research into all areas of biotechnology, including Bt crops.

The construction of Genome Valley was facilitated through aid funding from the UK government. This was despite the resistance to Bt crops in the UK and their ongoing ban due to safety concerns (Harding and Vidal 2001). The international support for the location of research into this globally controversial technology in India suggested that the potential for a 'legitimation crisis' (Habermas [1973], 1976) was passed to India, along with the funding.

As Frankel (2005: 616) highlighted, 'no part of the [*Vision 2020*] plan directly addresse[d] the structural inequalities in the state.' Nonetheless, proponents of the neoliberal approach and of Bt technology argued that both had significantly contributed to the alleviation of absolute poverty in the state. Between 2002 and 2012, the pre-secession Andhra Pradesh economy grew at an average rate of 8.2 per cent.[22] This was higher than the national growth rate of 6.5 per cent in 2011–12.[23]

Figures from the National Sample Survey reported in *The Hindu* indicated that the average poverty level in pre-secession Andhra Pradesh declined from 21 per cent to just 9 per cent between 2009–10 and 2011–12.[24] This purported decline was attributed by some economists to government initiatives such as the Indiramma Housing Initiative,[25] the Mahatma Gandhi National Rural Employment Guarantee Scheme (MGNREGS)[26] and higher wage rates for rural labour. The Public Distribution System (PDS), introduced in 1972, also served to alleviate rural risk through delivering subsidised food items to the most vulnerable.[27]

Other economists argued, however, that reports of a drastic reduction in poverty in the pre-secession state were exaggerated. They claimed that purported declines were due to issues of data comparison and sample bias.[28] Patnaik (2007: 3132) argued that nearly half of the rural poor were excluded from the official national Planning Commission estimate of 28 per cent for 2004–05. Patnaik (2007: 3132) claimed that rural poverty in India was, in fact, closer to 87 per cent.

Patnaik's view appeared to be supported by the finding of the 2011 Census that 95 per cent of the population of pre-secession Andhra Pradesh possessed white ration cards, associated with below poverty line (BPL) households, for supplies from the Public Distribution System.[29] Although there are noted issues with PDS targeting, the profusion of BPL ration cards in the pre-secession state suggested that rural poverty was more of an issue than Planning Commission figures indicated.

Galab et al. (2009: 191) argued that the situation of indebtedness of farmers in pre-secession Andhra Pradesh was worse than in other states. The *Times of India* (12/10/2010) reported that 93 per cent of Andhra Pradesh's rural poor were in debt. It was argued that this was linked to the raised expectations created by the *Vision 2020* initiative which led to increased pressure on consumption associated with, for instance, dowries and house construction (Rao and Suri 2006: 1552; Galab et al. 2009: 169; Deshpande and Shah 2010b: 129).

The ongoing privatisation arising from the liberalisation of the state saw increased private expenditure related to 'public' services, such as education and health (Galab et al. 2009: 170; Reddy 2010: 257). Such privatisation continues to contribute to a two-tier system in these services where the extremely vulnerable are confined to poorly resourced government hospitals and state-run schools.

The increased privatisation of health care is highlighted in the Srikrishna Report (Srikrishna et al. 2010: 171) which notes that public expenditure on health declined from 1.29 per cent of GSDP (Gross State Domestic Product) between 1985–90 to just 0.76 per cent by 2005–06. As the current research highlights, even some of the most vulnerable participants sought loans wherever possible to avoid exposure to the perceived risks of government (publicly-funded) hospitals.

Rao and Suri (2006: 1552) note that the defeat of Naidu's Telugu Desam Party in pre-secession Andhra Pradesh in 2004 was largely due to the delegitimation of his government by cultivators. Naidu's successor from the Congress Party, Y.S. Rajasekhara Reddy (locally known as YSR), introduced a wave of populist measures in an attempt to more strongly secure the government's legitimacy among farmers. None of these paid attention to the redistribution of resources; instead, they included policies such as the waiving of electricity charges for agricultural purposes, a moratorium on farmer debts and compensation payments to the families of farmers who had taken their lives due to agriculture-related issues (Rao and Suri 2006: 1552; Sridhar 2006: 1564).

Given the World Bank borrowing which had financed the investment in infrastructure associated with the *Vision 2020* initiative, the pre-secession government was caught between the demands of power- holders that conditionalities associated with liberalisation and globalisation be respected because this suited their own interests, while simultaneously being held to account by a strong NGO sector, radical Maoists[30] and a highly mobilised rural population. These groups continue to assert the risks of liberalisation, globalisation and a technocratic approach to agrarian poverty.

Despite its strong commitment to promoting a neoliberal, globalised approach which actively sought to secure the investment of multinationals, the government of the pre-secession state was nonetheless obliged to confront Monsanto in relation to its negotiation of Bt cotton as a concern for its own legitimacy. This was particularly given the scale of the mobilisations organised through an NGO sector which, as Stone (2011: 387) notes, has 'contested the new technological regime at every step.' It was this need to respond to the risk assertions of its citizens as a concern for its own legitimacy in a highly volatile context which led to the pre-secession Andhra Pradesh government being referred to as 'the most troublesome' in India (Jishnu 2010) with regard to Bt cotton, as referred to in Chap. 1.[31]

In 2005, the state government banned certain varieties of Bt cotton from the market due to their association with widespread crop failure.[32] In that same year, the Andhra Pradesh Agricultural Commissioner ordered Mahyco-Monsanto to pay compensation of Rs 10,000 per acre to farmers who had lost their Bt crop in Warangal, Nalgonda and Khammam districts.[33] Monsanto challenged the order in the Andhra Pradesh High Court, claiming that the government was 'harassing' them (Kumar 2010: 9).

The pre-secession state was also involved in a successful bid to force Monsanto to reduce its seed prices in 2006 from Rs 1800 per 425 gram packet (sufficient for one acre) to Rs 750 per packet.[34] Monsanto argued that the value of the trait fee which it charges is an intellectual property right whose regulation is outside the jurisdiction of the state government.[35] However, the pre-secession Andhra Pradesh government responded with legislation referred to as the Andhra Pradesh Cotton Seeds (Regulation of Supply, Distribution, Sale and Fixation of Sale Price) Act 2007. This asserted the power of the state to regulate the sale and pricing of Bt technology within its territory.[36]

The state legislation followed the deletion of cotton from the Essential Commodities listing by the central government in 2007. Given that commodities listed as essential require the issuing of trade licences by states, the removal of cotton limited the ability of state governments to regulate the prices of Bt cotton seeds. Many states responded by passing their own more stringent legislation as part of their constitutional right given that agriculture is stipulated as an area which falls within the federal state's jurisdiction. A release from the Press Information Bureau of the central government announced the re-inclusion of cotton on the Essential Commodities list in 2009 following a storm of protest from NGOs.[37]

The issue of Bt technology also saw the pre-secession state involved in conflict with the central government in Delhi. A proposed move by the Indian government to introduce a Biotechnology Regulatory Authority of India Bill (known as the BRAI Bill) which would involve less stringent regulation than that required in state legislation was strongly resisted by the pre-secession state government and activists alike (Kumar 2010: 10).[38] This was due to the recognised constraint this would impose on the state's ability to respond to the assertions of risk with regard to Bt technology by its own citizens.

Protests against Bt technology continued with the commercial approval of Bt brinjal, India's first food crop to receive such approval.[39] The issue of Bt brinjal has involved the Supreme Court within attempts to sue Monsanto for violating biopiracy laws in its development and for failing to obtain appropriate licences for field trials (Laursen 2012: 11). The case of Bt brinjal also continues to raise the issue of state legitimacy in relation to risk governance (Gupta 2011). As highlighted in Chap. 1, the mass mobilisations led to a nationwide consultation process held by the then Minister of State for the Environment and Forests, Jairam Ramesh. In February, 2010, he announced an indefinite moratorium on the cultivation of Bt brinjal pending the findings of future research.

In his statement announcing the moratorium, the minister claimed:

'Based on all the information…and when there is no clear consensus within the scientific community itself, where there is so much opposition from the state governments, when responsible civil society organizations and eminent scientists have raised many serious questions that cannot be answered satisfactorily, when the public sentiment is negative and when Bt brinjal will be the first genetically modified vegetable to be introduced anywhere in the world…, it is my duty to adopt a cautious, precautionary principle-based approach and impose a moratorium on the release of Bt brinjal.'

(Ramesh 2010: 16–17).[40] The decision recognises the centrality of the protection of citizens to the state's legitimacy. It has also contributed, however, to the ongoing legitimation struggle which surrounds Bt technology in India.

This chapter has explored the legitimation of democracy with regard to the Indian, and specifically, village context. It has highlighted the way in which the institutionalised practice of democracy in the villages, most notably associated with the electoral process and *gram sabha* meeting, is mediated through power relations in ways which influence the distribution of patronage and serve to reinforce the village power structure. The inter-village mobilisations organised by NGOs contribute to democratic legitimacy through operating beyond the institutionalised structure in order to problematise the power relations within it and to assert the right to justification of the vulnerable.

The chapter also explored the way in which the mobilisations concerning Bt technology represented a significant issue for the pre-secession state of Andhra Pradesh given its already precarious legitimacy in a highly volatile environment. The specific contexts of Telangana and Warangal, and the background to the volatility and vulnerability of the region where the study villages are located, will now be explored.

Notes

1. As mentioned in the previous chapter, Indira Gandhi served as Prime Minister of India over two terms for a total of 15 years (1966–1977 and 1980–1984). Between 1975 and 1977, she imposed Emergency rule on the country. During this time, democratic rights were suspended, and political opponents arrested. There was also a significant manipulation and by-passing of the institutions of government, with a subsequent damage to their legitimacy from which they have struggled to recover (Sharma 2010: 79).
2. During the research period, two protestors were killed by police during agitation against a power plant in Srikakulam in (pre-secession) Andhra Pradesh in 2010 (*The Times of India*, 15/07/2010).
3. Most notable during the research period was the high-profile scandal related to the 2G telecom scam and political corruption in the awarding of access services licences to telecommunications companies (*Deccan Chronicle*, 24/11/2010).
4. Nandi (2010: 37) notes that the word *goonda* denotes 'people given to lawless acts, prone to violence and active especially during tumultuous occasions, like riots.'
5. Available at: http://parliamentofindia.nic.in/ls/debates/vol2p3.htm. Accessed on 26/3/2017.

6. Available at: http://indiatoday.intoday.in/story/indian-democracy-today-is-stronger-in-the-minds-of-people-than-it-ever-was/1/283011.html. Accessed on 26/3/2017.

7. Forst (2007: 20) claims that justifications can be assessed for their normative validity in terms of 'whether they can be upheld reciprocally (i.e., without some of the addressees claiming certain privileges and without one's own needs or interests being projected onto others) and generally (i.e., without excluding the objections of anyone affected).' This book argues that such assessments are conducted through a process of legitimation.

8. In 2008, 250 farmers and consumers protested in Hyderabad, the state capital, against GM crops. This was organised by GM Free Andhra Pradesh. Available at: http://www.i-sis.org.uk/gmProtestsIndia.php. Accessed on 3/4/2017.

9. On 9th August, 2011, 15 states across India launched protests organised by farmers' unions and civic groups to protest against GM crops and the perceived corporate takeover of Indian agriculture. Available at: http://astm.lu/august-9th-declared-as-monsanto-quit-india-day/. Accessed on 26/3/2017.

10. The fact that resistance to GM crops extends beyond cultivators and the rural context is illustrated by the protest against GM crops involving 400 cities worldwide in the third global March against Monsanto in May, 2015. Available at: https://www.rt.com/news/261573-monsanto-global-protests-gmo/. Accessed on 26/3/2017.

11. Available at http://www.indianexpress.com/news/first-official-estimate-an-ngo-for-every-400-people-in-india/643302/. Accessed on 26/3/2017.

12. The PRI structure was inaugurated by Nehru in 1959 and revived in its second generation in 1978 by the Ashoka Mehta Committee. This followed an initially lacklustre implementation (Kumar 2006: 17). Following this revision, political parties were permitted to participate in *panchayat* elections (ibid.: 21).

13. At the next level up, the Mandal Parishad is responsible for the implementation of rural development programmes in association with the *panchayats*, co-operatives, voluntary organisations and other development institutions. And at the district level, the Zilla Parishad oversees and supports the performance of the Mandal Parishads through the collection of data, consolidation of development plans, supervision of budgets and distribution of funds. More details on the various duties associated with each level can be found in Jain (2006: 244–245).

14. This normative view of the role of the *gram sabha* with regard to risk assumes that villagers possess perfect self-awareness in relation to their relative risk exposure and that the most vulnerable are able to articulate this. This, of course, as the analysis explores is not the reality—hence, the significance of representation. This normative view also disregards village power relations and the role of village elites in manipulating the discourse on risk, an aspect which is explored in the research. Finally, the normative assumption that the state is capable of obtaining perfect information with regard to the risk exposure of each village, and has the political will, as well as the administrative infrastructure, to implement remedial action effectively, is problematic. This is due not only to the complexity of the context but also to the significant issues with political practice generally in India as noted in this chapter. The particular problems associated with the politics of the pre-secession state of Andhra Pradesh will be explored in Chap. 4.

15. Clause (6) of Section 243D of the 73rd Amendment noted that it was up to the discretion of individual states to make their own provisions concerning reservations for Backward Castes. Available at: http://india-code.nic.in/coiweb/amend/amend73.htm Accessed on 26/3/2017.

16. Kumar's (2006: 209) study explored *gram sabha* attendance in eight villages, two in each of the states of Madhya Pradesh, West Bengal, Maharashtra and Karnataka.

17. While the landless accounted for 61 per cent of attendees in the villages in Madhya Pradesh, they represented less than 20 per cent of the attendees in the villages in West Bengal, Maharashtra and Karnataka. Similarly, while 63 per cent of attendees in West Bengal were from the Scheduled Castes/Scheduled Tribes, upper castes dominated the meeting in Maharastra and Karnataka (75 and 68 per cent, respectively). This variability suggests support for the view expressed here that institutionalised democratic practice is determined by the way in which power is legitimated in the villages as a result of the numerical presence of particular castes, as well as their land-holding, and the approach to the exercise of power adopted by influential village leaders.

18. Toddy is an alcoholic beverage made from the sap of the palm tree.

19. The shift to a neoliberal approach in Andhra Pradesh followed structural adjustment undertaken as a result of borrowing from the International Monetary Fund (IMF) by the Indian government in 1991. The changed approach of both the central and federal governments as a result of conditionalities for IMF and World Bank borrowing involved the removal

of barriers to globalised trade and the increased withdrawal of the state from public spending on welfare measures (Le Mons Walker 2008: 573; Kohli 2009: 39).

20. Biotechnology is a broad area which incorporates the development of edible vaccines for cholera, rabies and hepatitis B, as well as gene therapies for various genetic disorders (Sharma 2004: 741). It is recognised, however, that the development of GM crops is by far the most controversial area of research in this diverse field (ibid.).

21. Hyderabad was the capital city of the state of Andhra Pradesh prior to secession. It is a globally significant IT and biotechnology hub, attracting considerable foreign direct investment (Rajan 2006: 77–103). Its significance is evident in the conflict involved in determining which state it should rightfully be assigned to post-secession. It was eventually agreed that the city would serve as a joint capital for Telangana and the divided state of Andhra Pradesh for a period of not more than 10 years.

22. Available at http://www.cess.ac.in/cesshome/pdf/draft_approach_to_12th_plan_for_discussion.pdf. p. 1 Accessed on 26/3/2017.

23. Available at: http://indiatoday.intoday.in/story/gdp-growth-sharply-down-at-6.5-per-cent-in-2011-12/1/198325.html. Accessed on 26/3/2017.

24. Available at: http://www.thehindu.com/news/national/andhra-pradesh/andhra-pradesh-leads-in-bringing-down-poverty/article4952012.ece. Accessed on 26/3/2017.

25. In 2006, the pre-secession Andhra Pradesh government launched the Indiramma (Integrated Novel Development in Rural Areas and Model Municipal Areas) initiative. This aimed to upgrade the houses of all of those who had ration cards and who lived in *kacha* houses (those made from mud with roofs of rice straw or other thatching materials).

26. At a national level, the Mahatma Gandhi National Rural Employment Guarantee Act (MGNREGA) was passed in 2006. The scheme is colloquially known as NREGS. It seeks to address rural risk exposure through providing at least 100 days of guaranteed wage employment to every household member who is capable of it, particularly when agricultural work is limited. The scheme has attracted criticism with claims of late payments and arguments that the increase in incomes in which it has resulted is leading to a rise in alcohol abuse among males in rural areas (*Deccan Chronicle*, 20/10/2010). Also: http://www.hindustantimes.com/india-news/nrega-benefits-are-mixed-oxford-study/article1-1146429.aspx. Accessed on 26/3/2017.

27. Under the terms of the Public Distribution System, households with an annual income of less than Rs 6000 are issued with white ration cards which entitle them to 5 kilograms of rice at two rupees per person per month from fair price shops. This is up to a ceiling of 20 kilograms. Pink card holders are entitled to rations at the same price up to a lower ceiling. Sugar, kerosene and cooking oil are also provided at subsidised rates. The PDS has been criticised for its inefficient targeting of the poor (Deb 2009: 70) and for its leakages and waste (Indrakanth 1997: 999). The PDS also notably requires access to cash, a commodity which is in very short supply among the extremely vulnerable.

28. Available at: http://www.thehindu.com/news/national/andhra-pradesh/andhra-pradesh-leads-in-bringing-down-poverty/article4952012.ece. Accessed on 26/3/2017.

29. Available at: http://indiatoday.intoday.in/story/95-per-cent-below-poverty-line-andhra-pradesh/1/164651.html. Accessed on 26/3/2017.

30. Communism in India is characterised by a number of different groups with varying ideologies and approaches to the state. Among the most revolutionary are the CPI (Marxist-Leninist) and CPI (Maoist) parties. The Communist Party India (M-L) was formed from a split in the Communist Party India (Marxist) in 1969 (Brass 1990: 299). The CPI (M-L) do not recognise the legitimacy of the state, arguing that it serves only as an instrument of 'compradore-bureaucrat capitalism' (Mohanty 1986: 253) and thus blocks a true democracy of the people. Party members seek agrarian revolution through violent class struggle and argue against US and Russian imperialism in India. The CPI (Maoist) party was founded in 2004 through the merger of the CPI (M-L) People's War and the Maoist Communist Centre of India (MCCI). Maoists are inspired by a Marxist-Leninist-Maoist ideology which aims to seize power from the state through protracted armed struggle. The CPI (Maoist) party is designated as a terrorist organisation in India under the Unlawful Activities (Prevention) Act. Given the particular presence of Maoists in Telangana, they will be discussed more fully in Chap. 4.

31. Available at: http://business.rediff.com/column/2010/apr/01/guest-bt-cotton-monsanto-is-back-in-courts-over-royalty.htm. Accessed on 27/3/2017.

32. Three varieties of Mahyco-Monsanto's Bt cotton seeds were involved – Mech 12, Mech 162 and Mech 184.

33. Available at: http://www.global-sisterhood-network.org/content/view/705/76/. Accessed on 27/3/2017.

34. Available at: http://business.rediff.com/column/2010/apr/01/guest-bt-cotton-monsanto-is-back-in-courts-over-royalty.htm. Accessed on 27/3/2017.
35. Available at: http://www.thehindubusinessline.com/todays-paper/tp-agri-biz-and-commodity/monsanto-ap-govt-cross-swords-over-royalty-payment/article1731571.ece?ref=archive. Accessed on 27/3/2017.
36. Details of the act can be found at http://faolex.fao.org/docs/pdf/ind119055.pdf. Accessed on 27/3/2017.
37. Available at: http://pib.nic.in/newsite/erelease.aspx?relid=54255. Accessed on 27/3/2017.
38. Available at: http://www.thehindu.com/todays-paper/tp-national/parliament-urged-to-pass-brai-seeds-bill/article2332299.ece. Accessed on 3/4/2017.
39. These protests have included fasts, mass rallies and public consultations. Available at: http://www.thehindu.com/sci-tech/agriculture/scientists-farmers-fast-to-protest-bt-brinjal/article97617.ece; http://www.i-sis.org.uk/gmProtestsIndia.php; http://www.esgindia.org/campaigns/press/say-no-bt-brinjal-say-no-release-genetic.html. All accessed on 27/3/2017.
40. Available at: http://webcache.googleusercontent.com/search?q=cache:http://www.moef.nic.in/downloads/public-information/minister_REPORT.pdf&gws_rd=cr&ei=0KjGWKudLtOogAbDmqjAAQ. Accessed on 27/3/2017.

Bibliography

Adeney, K., & Wyatt, A. (2010). *Contemporary India*. New York: Palgrave Macmillan.

Alam, J. (2004). What is happening inside Indian democracy? In R. Vora & S. Palshikar (Eds.), *Indian democracy: Meanings and practices*. London: Sage Publications.

Bandyopadhyay, D. (2001). Andhra Pradesh: Looking beyond 'vision 2020'. *Economic and Political Weekly, 36*(11), 900–903.

Brass, P. R. (1990). *The politics of India since independence*. New York: Cambridge University Press.

Bryld, E. (2001). Increasing participation in democratic institutions through decentralization: Empowering women and scheduled castes and tribes through panchayat raj in rural India. *Democratization, 8*(3), 149–172.

Buch, N. (2009). Reservation for women in panchayats: A sop in disguise? *Economic & Political Weekly, XLIV*(40), 8–10.

Chandhoke, N., & Priyadarshi, P. (2009). Introduction. In N. Chandhoke & P. Priyadarshi (Eds.), *Contemporary India: Economy, society, politics* (pp. vii–xvii). New Delhi: Dorling Kindersley (India).

Cohen, J. L., & Arato, A. (1992). *Civil society and political theory.* Cambridge: MIT Press.

Corbridge, S., & Harriss, J. (2000). *Reinventing India: Liberalization, Hindu nationalism and popular democracy.* Cambridge: Polity Press.

Corbridge, S., Harriss, J., & Jeffrey, C. (2013). *India today: Economy, politics and society.* Cambridge: Polity Press.

Deb, S. (2009). Public distribution of rice in Andhra Pradesh: Efficiency and reform options. *Economic and Political Weekly, XLIV*(51), 70–77.

Deshpande, R. (2004). Social movements in crisis? In R. Vora & S. Palshikar (Eds.), *Indian democracy: Meanings and practices* (pp. 379–409). New Delhi: Sage.

Deshpande, R. S., & Shah, K. (2010). Globalisation, agrarian crisis and farmers' suicides: Illusion and reality. In R. S. Deshpande & S. Arora (Eds.), *Agrarian crisis and farmer suicides* (pp. 118–148). New Delhi: Sage.

Forst, R. (2007). *The right to justification.* New York: Columbia University Press.

Forst, R. (2014a). *Justice, democracy and the right to justification.* London: Bloomsbury.

Forst, R. (2014b). *Justification and critique.* Cambridge: Polity Press.

Frankel, F. R. (2005). *India's political economy 1947–2004.* New Delhi: Oxford University Press.

Fraser, N. (2008). *Scales of justice: Reimagining political space in a globalizing world.* Cambridge: Polity Press.

Fricker, M. (2007). *Epistemic injustice: Power & the ethics of knowing.* Oxford: Oxford University Press.

Galab, S., Revathi, E., & Reddy, P. P. (2009). Farmers' suicides and unfolding agrarian crisis in Andhra Pradesh. In D. N. Reddy & S. Mishra (Eds.), *Agrarian crisis in India* (pp. 164–198). New Delhi: Oxford University Press.

Gandhi, M. K. (1930, February 13). *Young India.*

Gandhi, M. K. (1948, January 18). *Harijan.*

Gibson, C. (2012). Making redistributive direct democracy matter: Development and women's participation in the gram Sabhas of Kerala, India. *American Sociological Review, 77*(3), 409–434.

Gupta, A. (2011). An evolving science-society contract in India: The search for legitimacy in anticipatory risk governance. *Food Policy, 36,* 736–741.

Gupta, S. P. (2002). *Report of the commission on India: Vision 2020.* New Delhi: Planning Commission, Government of India.

Habermas, J. ([1973], 1976). *Legitimation crisis.* London: Heinemann Educational Books.

Habermas, J. (1996). *Between facts and norms.* Cambridge: Polity Press.

Habermas, J. (2008). *Between naturalism and religion.* Cambridge, UK: Polity Press.

Harding, L., & Vidal, J. (2001, July 7). This is the path to disaster: Clare short is in the hot seat for funding GM crops in India. *The Guardian.*

Haunss, S. (2007). Challenging legitimacy: Repertoires of contention, political claims-making, and collective action frames. In A. Hurrelmann, S. Schneider, & J. Steffek (Eds.), *Legitimacy in an age of global politics* (pp. 156–172). New York: Palgrave Macmillan.

Indrakanth, S. (1997). Coverage and leakages in PDS in Andhra Pradesh. *Economic and Political Weekly, 32*(19), 999–1001.

Jain, S. P. (2006). Panchayati raj in Andhra Pradesh yesterday, today and tomorrow. In S. B. Verma (Ed.), *Empowerment of the Panchayati raj institutions in India.* New Delhi: Sarup & Sons.

Kohli, A. (2009). *Democracy and development in India: From socialism to pro-business.* New Delhi: Oxford University Press.

Kothari, R. (2005). *Rethinking democracy.* New Delhi: Orient Blackswan.

Kulkarni, V. (2012). The making and unmaking of local democracy in an Indian village. *The Annals of the American Academy, 642,* 152–169.

Kumar, G. (2006). *Local democracy in India: interpreting decentralization.* New Delhi: Sage.

Kumar, P. (2009). *Panchayati raj institution in India.* New Delhi: Omega Publications.

Kumar, Y. V. A. (2010). *Seed bill 2010: An analytic view.* Centre for Sustainable Agriculture. http://agrariancrisis.in/wp-content/uploads/2012/08/SEED-BILL-2010-an-analytical-view.doc

Laursen, L. (2012). Monsanto to face biopiracy charges in India. *Nature Biotechnology, 30*(1), 11.

Le Mons Walker, K. (2008). Neoliberalism on the ground in rural India: Predatory growth, agrarian crisis, internal colonization, and the intensification of class struggle. *Journal of Peasant Studies, 35*(4), 557–620.

Lerche, J. (2008). Transnational advocacy networks and affirmative action for Dalits in India. *Development and Change, 39*(2), 239–261.

Locke, J. ([1690], 1967). *Two treatises of government.* Cambridge: Cambridge University Press.

Mandelbaum, D. G. (1970). *Society in India: Volume two continuity and change.* Bombay: Popular Prakashan.

Mehta, P. B. (2007). The rise of judicial sovereignty. *Journal of Democracy, 18*(2), 70–83.

Mohanty, M. (1986). Ideology and strategy of the communist movement in India. In T. Pantham & K. L. Deutsch (Eds.), *Political thought in modern India* (pp. 236–260). New Delhi: Sage.

Nandi, S. (2010). Constructing the criminal: Politics of social imaginary of the "goonda". *Social Scientist, 38*(3/4), 37–54.

Parthasarathi, G. (Ed.). (1989). *Jawaharlal Nehru: Letters to Chief Ministers Vol. 5 1958–64.* New Delhi: Jawaharlal Nehru Memorial Fund and Oxford University Press.

Patnaik, U. (2007). Neoliberalism and rural poverty in India. *Economic and Political Weekly, 42*(30), 3132–3150.

Powis, B. (2003). Grass roots politics and 'second wave of decentralisation' in Andhra Pradesh. *Economic and Political Weekly, 38*(26), 2617–2622.

Pur, K. A., & Moore, M. (2010). Ambiguous institutions: Traditional governance and local democracy in rural South India. *Journal of Development Studies, 46*(4), 603–623.

Rajan, K. S. (2006). *Biocapital: The constitution of postgenomic life.* London: Duke University Press.

Rao, P. N., & Suri, K. C. (2006). Dimensions of agrarian distress in Andhra Pradesh. *Economic and Political Weekly, 41*(16), 1546–1552.

Rao, V., & Sanyal, P. (2010). Dignity through discourse: Poverty and the culture of deliberation in Indian village democracies. *The Annals of the American Academy, 629*, 146–172.

Reddy, T. P. (2010). Distress and deceased in Andhra Pradesh. In R. S. Deshpande & S. Arora (Eds.), *Agrarian crisis and farmer suicides* (pp. 242–263). London: Sage.

Robinson, M. S. (1988). *Local politics: The law of the fishes: Development through political change in Medak district, Andhra Pradesh (South India).* Delhi: Oxford University Press.

Rousseau, J.-J. ([1762], 1973). *The social contract and discourses.* London: J.M. Dent & Sons.

Scanlon, T. M. (2012). Justification and legitimation: Comments on Sebastiano Maffettone's *Rawls: An introduction. Philosophy & Social Criticism, 38*(9), 887–892.

Sharma, D. C. (2004). Technologies for the people: A future in the making. *Futures, 36,* 733–744.

Sharma, S. (2010). Indian politics. In N. DeVotta (Ed.), *Understanding contemporary India* (pp. 67–94). London: Lynne Rienner.

Singh, M. P., & Verney, D. V. (2003). Challenges to India's centralised parliamentary federalism. *Publius, 33*(4), 1–20.

Sridhar, V. (2006). Why do farmers commit suicide? The case of Andhra Pradesh. *Economic and Political Weekly, 41*(16), 1559–1565.

Srikrishna, B. N., Duggal, V. K., Singh, R., Shariff, A., & Kaur, R. (2010). *Committee for consultations on the situation in Andhra Pradesh*. New Delhi: Government of India.

Srinivas, M. N. (1991). On living in a revolution. *Economic and Political Weekly, 26*(13), 834–836.

Stone, G. D. (2011). Field *versus* farm in Warangal: Bt cotton, higher yields, and larger questions. *World Development, 39*(3), 387–398.

Web-Sites

Jishnu, L. (2010). Bt cotton: Monsanto is back in courts over royalty, 1/4/2010. http://business.rediff.com/column/2010/apr/01/guest-bt-cotton-monsanto-is-back-in-courts-over-royalty.htm. Accessed 25 Mar 2017.

Shukla, A. (2010). First official estimate: An NGO for every 400 people in India. *The Indian Express*, 7/7/2010. http://www.indianexpress.com/news/first-official-estimate-an-ngo-for-every-400-people-in-india/643302/. Accessed 26 Mar 2017.

4

The Legitimation of Risk and Democracy in Telangana

This chapter explores the significance of the legitimation of risk and democracy in the context of the region of Telangana, and the district of Warangal within it, as the location for the research. It examines the historical struggle for legitimation associated with imperial rule in Telangana, as well as the unequal access to resources and acute exposure to risk with which it continues to be associated.

The chapter describes the way in which the perception of an illegitimate and unjust exercise of power by the pre-secession Andhra Pradesh government led to ongoing challenges from radical communists in the region, as well as to demands for secession by the Telangana movement. The chapter illustrates how the legitimation struggle which Bt technology represents in Telangana is occurring within a high-risk, highly politicised, volatile context, in which exposure to risk remains extremely differentiated. It argues that mobilisations concerning Bt technology seek to draw attention to the injustice of the inequality and risk exposure of those in the region, in much the same way as demands for secession did.

© The Author(s) 2018
E.L. Desmond, *Legitimation in a World at Risk*,
https://doi.org/10.1007/978-981-10-6065-6_4

Andhra Pradesh, Telangana and the Historical Struggle for Legitimation

The current research was conducted between June, 2010 and March, 2011 when the state of Andhra Pradesh still included the region of Telangana. This pre-secession state was located on the south-east coast of peninsular India. It comprised three Telugu-speaking regions—Telangana, Rayalaseema and Coastal Andhra. Pre-secession, Andhra Pradesh was the fourth largest of India's 28 states by area (275,000 square kilometres) and fifth by population (84 million inhabitants).[1] Sixty-six per cent of the population lived in rural areas, and the state was home to 28,123 villages.[2] The climate was noted for its extreme variability. In the 33 years from 1977 to 2010, the Disaster Management Department of the pre-secession government reported 55 natural disasters related to flooding, droughts and cyclones.[3]

Srinivasulu (2002: 4) notes that the pre-secession state was dominated by a large variety of Backward Castes who comprised 46 per cent of the population. Forward Castes accounted for 29 per cent and, of these, the Reddys were among the most powerful (Wade 1994: 31). Despite the dominance of the Reddys, however, they accounted for just 6.5 per cent of the population (Srinivasulu and Sarangi 1999: 2449).

A number of other Forward Castes, most notably the Kammas, Kapus and Velamas, struggled for land and political power. The Srikrishna Report (Srikrishna et al. 2010: 380) highlights that the Reddys and Kammas held economic and political power in the pre-secession state, while the Velamas and Kapus struggled to improve their position. As Gundimeda (2009: 52) notes, along with the Backward Caste Gowdas, these castes were 'locked in a fierce caste-war' for political and economic power in the pre-secession state. There was only a small number of Brahmins—just 3 per cent of the total population (Srinivasulu 2002: 4).

The Scheduled Caste (SC) grouping (those formerly designated as *Dalits* or 'Untouchables') is dominated by two castes in Telangana and Andhra Pradesh—the Madigas and Malas. In the pre-secession state, Scheduled Castes accounted for 16 per cent of the total population

(Srikrishna et al. 2010: 63). Of this, the Malas accounted for 42 per cent of the SC population, while Madigas comprised 49 per cent (ibid.: 368). In Telangana, however, Madigas account for 61 per cent of the SC population (ibid.: 369).

Omvedt (1994: 90) notes that Malas are ritually higher than Madigas, and that Madigas are poorer and more of them are landless. Recent assertions of their particular deprivation by the Madigas have seen them demand a sub-categorisation of the SC category, a claim which was rejected by the Supreme Court (Srikrishna et al. 2010: 63). Srinivasulu (2002: 4) highlights that the Scheduled Castes formed the bulk of the agricultural labour in the pre-secession state. Scheduled Tribes account for 10 per cent of the Telangana population, a larger proportion than is found in any other region, and Muslims account for 8.4 per cent (Srikrishna et al. 2010: 115).

Jangam (2016: 25) highlights how *Dalits* became allied with wealthy Kamma cultivators against Brahmin and Reddy caste leaders given the involvement of both groups in the communist party. As Jangam (ibid.) notes, *Dalits* and agricultural labourers are strongly mobilised in Telangana and are particularly drawn to communism given their historic marginalisation.

Following decades of struggle, the Telangana region where the three villages in this study are located was officially declared as India's twenty-ninth state on 2nd June, 2014. Telangana is an inland semi-arid region. It has a population of 35 million people[4] and lies on the Deccan Plateau, an elevated area which covers much of south India. The plateau is spanned by the mountain ranges of the Western and Eastern Ghats. This book argues that the concerns asserted during the research period by those demanding secession in Telangana are reflected in protests against Bt cotton and that both can be seen as part of the ongoing struggle involved in the legitimation of risk and democracy in the region.

In order to understand the struggle for legitimation which Bt cotton represents in Telangana, it is necessary to appreciate the particular context in which the adoption of the technology has occurred. This recognises Bernstein's (2004: 18) view that understandings of legitimacy are 'highly contextual, based on historical understandings…and the shared norms of the particular community granting authority.' These norms and understandings have themselves emerged from past legitimation struggles, and challenges to these understandings will form part of ongoing legitimation struggles into the future. In the case of Telangana, the concern with legitimacy can be traced to the region's long centuries of imperial rule.

Mulligan (2007: 84) argues that 'the goal of so much political thought, for centuries, has been to discover, imagine, or create the condition of legitimacy.' The 'unceasing wars' (Inden 1990: 185) of empire associated with changes in dynastic rule throughout India's long history can be seen as resulting from challenges to the legitimacy of rulers within the attempt to secure a more just exercise of power and access to resources in the negotiation of risk.

It was during the Kakatiya Empire (1083–1323), with the city of Warangal as its capital, that a distinctive Telugu culture began to emerge in the Telangana, Coastal Andhra and Rayalaseema regions which would later form the original state of Andhra Pradesh. In 1687, Warangal was captured by the Mughal Emperor, Aurangzeb, from Delhi (Cohen 2010: 48; Metcalf and Metcalf 2012: 23). Aurangzeb established Hyderabad State in 1724 and, by 1725, the Nizam-ul-Mulk, the emperor's viceroy, had become the independent ruler of the 'Mughal Deccan' (Kakar 1995: 11; Mehta 2005: 144).[5]

Throughout British rule, the Nizam remained as de facto ruler of Hyderabad State, although some territory was yielded to the British (Ramakrishna 1983: 1–2). This surrendered territory included the districts of Coastal Andhra (known as the Circars) in 1765[6] and, in 1800, the districts of Rayalaseema (referred to as the Ceded Districts). The British invested significantly in the development of the fertile Coastal Andhra region, particularly with regard to irrigation.[7] Meanwhile, exposure to risk in Telangana under the Nizam's rule was acute and highly differentiated.

The state of Hyderabad was predominantly agrarian (Vaikuntham 2004: 98). Agriculture was beset with difficulties, however, and drought and famine were common (ibid.: 115). Srinivasulu (2002: 6) notes that risk in the state was exacerbated by a class of large land-holders (Muslim *jagirdars* and Hindu (upper caste Reddy and Brahmin) *deshmukhs*). These land-holders represented the Nizam's power base. They inflicted severe suffering on the local population through illegal evictions, the extraction of free goods and services (known as *vetti*) and the denial of people's dignity and self-respect (Srinivasulu 2002: 6).

Fifty per cent of agricultural produce was collected as revenue under a *diwani* system which accounted for 60 per cent of the land (Robinson 1988: 49).[8] Of the remaining land, 30 per cent was controlled by *jagirdars* and *deshmukhs*, and 10 per cent was kept for the maintenance of the Nizam (Robinson 1988: 49; Vaikuntham 2004: 98). As Roosa (2001: 58) highlights, the Nizam's state 'squeezed taxes out of the 10,300 villages of Telangana [and] invested very little back into them.' Meanwhile, the last sultan of Hyderabad, Nizam Mir Usman Ali Khan (r. 1911–48), was one of the world's wealthiest individuals (ibid.).

The communist-led Telangana Armed Struggle for which the region is renowned sought to challenge the legitimacy of the Nizam's rule and the unequal access to key resources which it supported.[9] Ravi (2015: 115) observes that the struggle began in the Warangal district. It was caused by the killing of Doddi Komarayya, a member of the Andhra Mahasabha, a people's organisation. This killing occurred during a struggle with the workers of a Reddy *deshmukh* which sought to prevent the landlord from seizing the harvest of Ailamma, a washerwoman (Sundarayya 1972: 17; Roosa 2001: 66).

The ensuing revolt soon extended to other areas of exploitation in the region, most notably the injustices associated with the grain levy collection (Roosa 2001: 63). Wealthier villagers would bribe collectors not to collect their grain. This meant that the collector's quota had to be met through additional extractions from more vulnerable villagers who could least afford to spare them.

By 1948, with the support of the communists, the rebellion broadened to a more violent concern with the redistribution of land, and the rural elite fled to the cities. The erosion of the class structure at this time heightened the awareness of the poor that their exposure to risk was neither natural nor unalterable but was a direct result of the Nizam's rule and their exploitation by intermediaries. Thus, as Roosa (2001: 77) reports:

[T]he village poor...realised that they [could] have their stomach full if they [did] not pay rents, taxes, debts, and if they divide[d] the land of the landlords and [fought] for higher wages.

Even during the rebellion, however, there was differentiation in the way in which captured lands were distributed. Srinivasulu (2002: 6) notes that 'while the lands of [the] *doras* [a local Telangana term for a large land-holder] were distributed among the [upper caste]…Reddy [cultivators] and tenants, the common pastures and waste lands became the lot of the landless *Dalits* and other lower castes.'

The brutality of the Nizam's armed militia, known as *Razakars*, meant that the rebellion turned increasingly violent. Roosa (2001: 81) notes that the Andhra Provincial Committee responsible for Telangana was controlled by radical communists who advocated 'prolonged civil war in the form of agrarian revolution, culminating in the capture of political power by a Democratic front.' Ravi (2015: 113) claims that this involved the communists in taking up 'day-to-day issues of the people' and resulted in a 'new awakening [which] contributed to anti-zamindari and anti-feudal struggles' (ibid.). By 1948, 3000 villages were involved, mostly in the districts of Warangal, Nalgonda and Khammam (Sundarayya 1972: 8).

While the redistribution of resources was welcomed by the rural poor, the power of the communists and the violence in the region became a concern for Nehru's Congress government in the newly independent India. As Roosa (2001: 80) notes, Nehru's decision to send the Indian army into Hyderabad in 1948 sought not only to depose the Nizam but also to confront the communists whose violent methods were perceived as a threat to the legitimacy of the federal union of India.

Nehru's efforts to establish the legitimacy of Delhi's central government in Telangana involved attempts at nonviolent land reform. The Nizam's lands were taken over for redistribution by the state (Agnihotri and Subramania 2002: 12). There was also a decline in *dora* dominance in the region given that many large land-owners were disinclined to reside in the villages after the struggle, and so disposed of their lands to former tenants (Srinivasulu 2002: 6). Meanwhile, the Bhoodan movement sought to persuade wealthy land-owners of their moral duty to give some of their lands to the poor.[10]

The Human Development Report compiled by the Centre for Economic and Social Studies referred to previously (CESS 2008: 65) outlines the impact of the Andhra Pradesh Land Reforms (Ceiling on Agricultural Holdings) Acts, passed in 1961 and 1973. The report (ibid.)

states that, while the 1961 legislation failed, that in 1973 had some limited effect. It claims that, of an estimated 2 million surplus acres in the pre-secession state of Andhra Pradesh in 1973, 790,000 acres were declared as surplus. Of this, 647,000 acres were claimed by the government and, of this, 582,000 acres were distributed to the poor and landless. By the end of 2002, 4.3 million acres of government land had been assigned to 2.3 million beneficiaries of whom 24 per cent were Scheduled Castes and 28 per cent were Scheduled Tribes.[11] The CESS report (ibid.: 29) notes, however, that much of this redistributed land was uncultivable because landlords handed over their least viable holdings.

Nonetheless, as a result of these changes, the proportion of small and marginal land-holders (i.e. those with less than 5 acres) in Telangana rose from 66 per cent of land-holders in 1970–71 to 82 per cent by 1995–96, and the proportion of land cultivated by these categories of landowner rose from 19 per cent to 45 per cent in the same period (Venkateshwarlu and Srinivas 2000: 10). The size of land-holdings also declined. In 2010–11, the average land-holding in Telangana was 2.8 acres (1.12 hectares) as opposed to an all India average of 1.16 hectares.[12]

Despite these changes, significant inequality remains. By 2008–09, 5 per cent of cultivators in Telangana owned 32 per cent of the land, and 44 per cent of the population were landless (Reddy et al. 2012: 57).[13] The inequality in the region is also noted in the Srikrishna Report (Srikrishna et al. 2010: 110) which found that inequality in Telangana had increased since 1993. The report notes that the differential was particularly great among Scheduled Castes and Tribes, as well as Muslims, and was highlighted as an 'important indicator of the unrest between communities' (ibid.: 119).

The concern with access to resources as the means to negotiating risk was central to the debate preceding Telangana's amalgamation—along with Coastal Andhra and Rayalaseema—into the wider state of Andhra Pradesh in 1956.[14] While supporters of the merged state argued that Telangana would benefit from a Coastal Andhra food surplus and increased investment through being part of a larger state (Ali et al. 1955: 104), opponents asserted Telangana's particular vulnerability as a result of its exploitation under the Nizam's rule.

Melkote et al. (2010: 9) claimed that Telangana was rich in resources but 'lagged behind due to the absence of English educational facilities and employment opportunities under the Nizam.' This led to the fear that 'Telangana [would] become a colony of the Andhra region' (Haragopal 2010: 52). It was argued that the region's incorporation into the larger state would result in its marginalisation and limit its ability to gain recognition and representation in the competition for resources. While a majority in Hyderabad's Legislative Assembly supported a united state, many continued to doubt its legitimacy (Rao 1988: 166).

The 'Gentleman's Agreement' (Brass 1990: 236) which eventually secured the amalgamation of Telangana into Andhra Pradesh involved reassurances that the interests of Telangana locals or *mulkis* would be protected. This included guarantees that one-third of places in public services, the state ministry and educational institutions would be reserved for *mulkis* (Brass 1990: 234; Melkote et al. 2010: 9). It was also stipulated that the Chief Minister and Deputy Chief Minister would be from different regions in order to ensure balanced representation in the state (Srinivasulu 2002: 7).

According to Maringanti (2010: 35), however, these promises 'remained largely on paper.' In terms of representation, the first Chief Minister of Andhra Pradesh refused to name a Deputy Chief Minister from Telangana, arguing that the Deputy Chief Minister-ship was like the 'unwanted sixth finger of a hand' (Rao 1988: 186). This inadequate representation led to a more general failure to keep to the terms of the agreement in ways which, it was claimed, severely impacted upon the ability of those in the region to combat risk.

It was argued, for instance, that those in Coastal Andhra had obtained false *mulki* certificates which 'enabled them to settle down in Telangana and to buy vast stretches of land' (Melkote et al. 2010: 8). This contributed to higher land prices and weakened the ability of the landless in Telangana to gain access to land (Kannabiran et al. 2010: 74).

In the case of employment, it is recognised that the ability of household members to secure an additional income in the non-farm sector represents a key means to mitigating agrarian risk. This is due to the fact that such employment not only reduces the number of family members who are reliant on a precarious agrarian existence; it also permits individuals to alleviate the risk of their family through sending regular remittances to supplement variable farm incomes.

Melkote et al. (2010: 8) note that, even prior to the formation of Andhra Pradesh, many from Coastal Andhra had begun to obtain employment in Telangana as their English education and experience in British administrative procedures gave them an advantage over local people. This, it was argued, was reducing the employment options for Telangana locals, particularly in the case of the much sought after jobs in government (ibid.).

The legitimation struggle arising from the breaches of the Gentlemen's Agreement had become particularly heated by the time of the current research. This book argues that these protests demanding secession formed part of a historic struggle for legitimacy in Telangana which also informs the protests against Bt cotton.

The contribution of the volatility in Warangal to the way in which the controversy concerning Bt cotton is negotiated in the district is highlighed by Herring. According to Herring (2008: 151), Mahyco-Monsanto was forced to pay Warangal farmers Rs 30 million in compensation as a result of the crop failure linked to the withdrawn varieties in the pre-secession state.[15] As Herring (ibid.: 150) notes, Warangal remains the only district in India where compensation for crop failure has been awarded. Herring (ibid.) argues that this is due to the sensitivity of the district's administration to 'rural political protest because of [the] history of Naxalite activity,'[16] as well as the district's NGO sector (ibid.: 155). This book argues that the legitimation struggle concerning Bt technology seeks to gain recognition and representation for the wider context of risk within which the issue of Bt cotton is embedded as the basis for asserting the right to justification of the vulnerable. This will now be explored in relation to Telangana and Warangal.

Bt Cotton in Telangana and Warangal

Sixty-one per cent of Telangana's population live in rural areas and the region comprises 10,128 inhabited villages.[17] The district administration for Telangana is based in Warangal city, which has a population of 615,998 people.[18] As highlighted previously, the extent of the ongoing and highly differentiated exposure to risk in Telangana is evident from the disproportionately high numbers of farmer suicides with which Telangana, and the district of Warangal in particular, has been associated.

Given the high occurrence of farmer suicides in Telangana, there has been a great deal of focus on trying to find reasons for them. As discussed in Chap. 1, a number of studies have found that indebtedness is an important factor, even while the reasons for such indebtedness are noted as being multiple and varied. Attention has also been given to the sources of debt. Iyer and Arora (2010: 279), for instance, note that the risks of indebtedness are exacerbated through accessing credit from non-institutional sources, such as money lenders. This was particularly evident during the current study in the controversy surrounding Micro-Finance Institutions (MFIs).[19]

The coercive attempts of MFIs to recover loan repayments in the midst of the catastrophic 2010/2011 season in which the research was undertaken, which was associated with widespread flooding, resulted in farmer suicides (*Times of India*, 17/10/2010). It was reported that 30 farmers took their lives in 45 days in the pre-secession state as a direct result of the approach adopted by MFIs (ibid.). These included three farmer suicides in 2 months in Warangal (*Times of India*, 13/10/2010).

The experience of MFIs is in stark contrast to Self-Help Groups (SHGs). SHGs are group savings schemes which are predominantly associated with females (Galab and Rao 2003; Tesoriero 2005). These groups alleviate some of the risk associated with borrowing given the lower interest rates and more personal approach which they seek to foster. SHGs have been linked to female empowerment (Tesoriero 2005: 329). In the current study, the organisation of SHGs in the villages coincides with the presence of NGOs.

It should be noted that the Telangana region is associated with particular risk factors with regard to cultivation, regardless of crop choice. It is drought-prone; hence, access to water is a crucial factor in the negotiation of agrarian risk. Vakulabharanam (2004: 1421) claims that the gradual withdrawal of state support for irrigation as part of the liberalisation reforms in pre-secession Andhra Pradesh led to the expansion of borewell irrigation in the region. This was largely funded through credit.

Irrigation through private spending on borewells in Telangana increased by 44 per cent between 1997 and 2002 (Reddy 2010: 245). It is recognised that this dependence on ground-water as a means of irrigation is unsustainable due to a declining water table (Galab et al. 2009: 175).[20] A citizen's report (CESS 1998: as cited in Galab et al. 2009: 168) which investigated 50 households of deceased cotton farmers in the Warangal district found that, in all cases, suicides coincided with crop failure due to inadequate water supply.

Agriculture in the pre-secession state had also become increasingly non-viable due to rising costs. Reddy (2010: 246) claims that cultivation costs increased by 50 per cent in Andhra Pradesh during the 1990s. Sridhar (2006: 1563) also asserts that, prior to secession, the prices of inputs in the state were among the highest in the country, not least due to the fact that input suppliers also provide credit to farmers, a power relation which often obliged farmers to accept higher prices.

It was argued that these increasing costs were not being off-set by higher returns. With regard to cotton cultivation, Galab et al. (2009: 181) note that, while incomes and yields associated with cotton grew until the mid-1990s, these had been declining. It was also noted that cotton cultivators were exposed to the risks of a speculative global commodity market over which they had no control (Rao and Suri 2006: 1546). Along with yield uncertainty, variations in output prices contributed to a high degree of volatility in farmer incomes (Galab et al. 2009: 183). Deshmukh (2010: 182) also highlighted the impact of power relations on risk negotiation with regard to prices obtained for crops, claiming that influential farmers were offered preferential prices.

As highlighted in Chap. 2, according to Revathi's (2009: 217) analysis, farmer suicides have traditionally been associated with small and marginal farmers. Galab et al. (2009: 191) note that the risk of non-institutional credit is particularly linked to these categories of land-holding given 'an inverse relationship between the size of land-holding and access to institutional credit.'

The finding that the risks associated with the agrarian crisis are generally more acute for small and marginal land-holders is also asserted by

Galab and Revathi (2009: 187). These authors found that, according to income and expenditure levels in 2002–03, agriculture was viable only for cultivators with holdings of 10 acres (4 hectares) or more.[21] Access to land has been increasingly constrained, however, by the reduction in the net sown area in Telangana and an increase in fallow land (Galab et al. 2009: 173).

The Human Development report on Andhra Pradesh (CESS 2008: 67) noted a decline in the net sown area in eight districts of the pre-secession state, six of which were in Telangana. (According to this report, the decrease in net sown area in the district of Warangal was relatively low). The report (ibid.: 76) asserts the need for the reasons for this decline to be identified at local levels and steps taken to redress it.

Galab et al. (2009: 173) argue that land is often left fallow due to its degradation which they claimed had affected 19 per cent of the cultivated area of the pre-secession state. Galab et al. (ibid.) asserted that much of this degradation had resulted from the stress on soils arising from unsustainable agriculture practices in drought-prone areas, such as Telangana. These practices include aspects such as a failure to undertake crop rotation and mixed cropping, as well as the inappropriate use (and over-use) of chemical fertilisers and pesticides.

Attempts by small farmers to enhance the viability of their cultivation by supplementing their access to land through leasing heighten their exposure to risk in the event of crop failure. This is due to the fact that the land lease rent is payable even if the crop fails. A study conducted by Reddy (2010: 250) in four districts of pre-secession Andhra Pradesh, including Warangal, found that 25 per cent of farmers who committed suicide were tenants.

Reddy (2010: 254) highlights that there are wide variations in leasing arrangements in Telangana which are often agreed verbally between landlord and tenant. In his 2004 study in four districts of the pre-secession state which included Warangal, Reddy (ibid.) notes the popularity of 'sharecropping' where the tenant and landowner shared input costs and profits equally. However, he also notes arrangements where an annual rent, varying between Rs 500 and Rs 2500 per acre, depending on the availability of irrigation, was payable.

In the current study, costs of the land lease rent varied between Rs 2000 and Rs 5000 per acre, depending on access to irrigation. There was evidence of arrangements where profits and costs were shared equally in addition to the rent, sometimes only the profit and costs were shared in the absence of rent or, sometimes, only the rent itself was payable.[22]

Galab et al. (2009: 193) argue that the higher incidence of suicides among small and marginal farmers is associated with their 'moving from subsistence agriculture to… high-value crops with a strong motivation to improve their social and economic status.' This has been referred to as 'the tragedy of the upwardly mobile' (Rao 2009: 109). It is particularly associated with the 'revolution of rising aspirations' (Gupta 2002: 86) created by *Vision 2020* and the liberalisation of the economy in the pre-secession state discussed in Chap. 3.

Fieldwork in all three villages found that the Indiramma initiative, which aimed at relieving poverty, had often contributed to borrowing in order to build larger, more individualised homes than the government funding permitted. As Revathi and Galab (2010: 201) note, the increased aspirations of cultivators in states with high economic growth coincided with higher numbers of farmer suicides. The current study suggests that these heightened aspirations are linked to indebtedness and the legitimation of the taking of financial risks as the means to development.

The district of Warangal where the three villages are located has a population of 3.5 million people, of which 72 per cent live in rural areas.[23] Herring (2008: 150) highlights that the risk of cotton cultivation in the district is particularly high given that it is drought-prone and cotton is often grown on unirrigated, thin red soils.[24] Nevertheless, Warangal is particularly dependent on cotton, and the crop is cultivated on 26 per cent of the cultivated area.[25] This makes it the second largest crop after paddy (which accounts for 32 per cent).[26] This reliance on cotton has been linked to the high numbers of farmer suicides in the district (Stone 2011: 390).

Herring (2008: 148) notes that Warangal's Department of Agriculture estimates Bt cotton adoption in the district to be close to 95 per cent. The pervasiveness of Bt cotton also coincides, however, with significant opposition to it.[27] This is led by an NGO sector in a Warangal district which, as Herring (2008: 155) observes, appears to be 'especially densely populated by NGOs.'

Gaurav and Mishra (2012: 14) assert that the additional costs of Bt cotton heighten the risk exposure of cultivators in an already high-risk context and are not always off-set by higher yields. This is particularly the case in the event of crop loss, a risk which is particularly acute in Telangana given the vagaries of the climate.[28] Because of the high costs involved in Bt cotton adoption, NGOs advocate alternative lower cost cultivation practices, most notably Non-Pesticide Management (NPM) and organic practices.

NPM methods are pioneered by the Centre for Sustainable Agriculture (CSA), in conjunction with a variety of local NGOs.[29] NPM practitioners treat pests through natural solutions which, it is claimed, eliminate the need for expensive chemical pesticides.[30] Ramanjaneyulu and Kuruganti (2006: 563), claim that, by 2006, NPM methods were being used on more than 10,000 acres across different districts in pre-secession Andhra Pradesh. This accounted for just over 0.05 per cent of the total cultivated area. Misra (2009: 22) states that, of the 3000 NPM villages in the pre-secession state in 2007, not a single suicide death was registered. Despite this, however, Ramanjaneyulu and Kuruganti (2006: 563) argue that the pre-secession government continued to focus on Bt cotton without providing adequate support for alternative agricultural practices, such as NPM.[31]

In the case of organic farming, Eyhorn (2007: 29) claims that, by 2005, organic cotton projects in pre-secession Andhra Pradesh, promoted by the NGOs, Chetna and Oxfam, had been initiated on 2400 acres. This accounted for approximately 0.01 per cent of the cultivated area in the pre-secession state. Both NPM and organic methods use non-Bt seeds, bulk orders for which have to be coordinated by NGOs given the problems with obtaining non-Bt seeds from seed dealers in Warangal. The latter difficulty was also noted by Stone (2011: 390).

Sales of spurious Bt cotton seeds exacerbate the risk of crop loss associated with Bt cotton. Herring (2008: 151) notes that spurious seeds are sold by unscrupulous dealers as Bt seeds when they do not, in fact, contain the Bt gene. Alternatively, they may be F2 second generation seeds gathered from the seed saving of a Bt cotton crop in which the genetic modification is associated with a weaker expression (ibid.). The issue of spurious seeds led to widespread suicides in Warangal in 1997–98

(Sridhar 2006: 1562). It also resulted in protests which involved the burning of seed outlets in Warangal city (Qayum and Sakkhari 2005: 5).

There have also been reports of animal deaths linked to Bt cotton. Kuruganti (2006: 4246) argues that such deaths indicate serious short-comings in the biosafety testing in India, given the failure to test the impact of Bt crops on livestock. As highlighted, Herring (2008: 155) describes the animal deaths as 'biologically impossible.' Nonetheless, reports of such deaths from a number of villages led to the issuing of an advice by the then Director of Animal Husbandry in Warangal that farmers should not permit their animals to graze on the crop. In a telephone interview with the former Director who issued the advice (17/2/2011), he asserted that his recommendation was offered as a precautionary measure only as a response to NGO protests. In his view, the sporadic reports of such deaths meant that they were more likely to be due to pesticide residues rather than the Bt cotton plant itself.

The issue of animal deaths has become central to the legitimation struggle associated with Bt technology in the villages. This has increasingly problematised the way in which the perspectives of cultivators are recognised and represented within the wider democratic process with regard to decision-making on risk. The concern of Bt cotton protests with gaining recognition and representation for risk, and the highly unequal exposure to it, was echoed in the demands for secession. This book argues that mobilisations for both state secession and Bt cotton sought to secure a more legitimate approach to risk and to democratic practice, a claim which will now be explored with regard to the Telangana movement.

The Demand for State Secession in Telangana

Michelutti (2007: 639) notes that 'values and practices of democracy become embedded in particular cultural and social practices, and… entrenched in the consciousness of ordinary people.' This 'vernacularisation' (ibid.) of democracy results in the adoption of different methods to challenge injustice through local processes of legitimation and contributes to differences in what is perceived as legitimate within particular societies. It is argued here that the struggle for the secession of Telangana represented an effort to secure the legitimation of risk and democracy as part of an ongoing concern with justice in the exercise of power and allocation of resources in response to the risk of the region.

The demand for secession was led by the Telangana movement.[32] Renn (2008: 135) argues that 'risk issues have become battlegrounds for substantial conflicts about resource allocation, social justice and future economic development.' This book claims that these 'battlegrounds' represent legitimation struggles. The centrality of the concern for legitimation in the fight for secession involved the assertion of the right to justification for the existing allocation of resources and exercise of political power. The region's eventual secession illustrated Clark's (2007: 195) view that 'legitimacy [is] a constitutive element within a society.'

Those in the Telangana movement argued that the pre-secession Andhra Pradesh government was 'dominated by people from the [Coastal] Andhra region' (Brass 1990: 234). This, they claimed, meant that 'more money and attention was given [to Coastal Andhra] by the state government' (ibid.). The linking of the allocation of resources required to negotiate risk with democratic legitimacy highlights the inextricable nature of attempts to secure the legitimation of risk and of democracy which this book asserts. The implicit recognition of the link between these two aspects of legitimation underpinned the way in which demands for a separate state were framed.

Kannabiran et al. (2010: 79) note that the appeal for a separate state served to 'nurture deliberative politics' and to encapsulate 'the demand for justice and equality, for freedom from want, for security, for a meaningful life' (ibid.: 80). Here, the formulation of the right to justification as part of the demand for secession involved deliberation and reflexivity on democratic practice itself as a concern for enhancing its legitimacy with regard to the negotiation of risk. These demands for greater democratic legitimacy in the exercise of state power encapsulated concerns related to all three of Fraser's (2008: 16–17) dimensions of justice— namely, redistribution, recognition and representation.

In terms of redistribution, the concern for the unequal allocation of resources, most notably between Coastal Andhra and Telangana, led to the refrain of '*maadi maaku kaavale*' ('we want what is ours') among supporters of a separate state (Kannabiran et al. 2010: 69). It was asserted that, although revenue from the Telangana region (excluding Hyderabad) formed more than half of the total income for the pre-secession state between 2003–04 and 2006–07, the region was not receiving its fair share of resources (Melkote et al. 2010: 9).

This was particularly evident with regard to irrigation.[33] Protestors argued that the failure to provide adequate water to the region encouraged irrigation through private borewells, the expense of which heightened the vulnerability of Telangana cultivators to indebtedness and suicide (Kannabiran et al. 2010: 72; Melkote et al. 2010: 10).

It was claimed that the same was true with regard to employment. As noted in the Srikrishna Report (Srikrishna et al. 2010: 162–163), many women tended to favour a separate state in the belief that it would bring their children education and jobs. In the case of those benefiting from reserved places in education, these were often first generation college graduates in their families whose expectations had been heightened through their access to education. The Srikrishna Report (ibid.: 162) notes, however, that the provision of jobs did not keep up with the supply of graduates which added to the frustration of Telangana's youth.

In relation to recognition, Kannabiran et al. (2010: 78) claimed that, prior to secession, the Andhra Pradesh government's engagement with the Telangana movement had involved the adoption of 'hegemonic violence' as a means of stifling protests. Maringanti (2010: 36) argued that the pre-secession government 'branded even the most direct demands for grievance redressal in villages as militant activity and dealt with it through police repression.' This authoritarian approach contributed to the perception of many in Telangana that the powerful from Coastal Andhra had formed a 'ruling nexus of contractors, politicians and businessmen' (Maringanti 2010: 34) which was constraining their ability to gain recognition for their inadequate access to resources and heightened exposure to risk.

The significance of representation lay in the belief that the demands for the right to justification of those in Telangana could best be met through a smaller state. The latter would, it was argued, permit more direct representation in the central government in Delhi, rather than having the risk exposure of those in Telangana represented through a power structure which was itself charged with their exploitation.

Maringanti (2010: 34) noted that the feeling among protestors was that the 'prospects for a better life [in Telangana] lay only in its gaining a bargaining position in the federal structure of the Indian state.' Given the composition of Telangana's population, the lower castes and minorities also felt that they would be in a much stronger position with regard to their recognition and representation in a smaller state because their proportion of the total population would be greater (Srikrishna et al. 2010: 390).

The perceived illegitimacy of state power in pre-secession Telangana contributed to the legitimation of radical communist activity in the region in the form of Naxalism. Naxalites seek to disrupt power relations through the violent redistribution of land and their support for agricultural labourers in conflicts with large land-holders.

The presence of Naxalites in Telangana can be seen as a remnant of the armed struggle. Their legitimation by many of the most vulnerable arose from the perceived failure of the pre-secession government to more adequately address the historic inequality in access to resources, as well as the ongoing acute and differentiated exposure to risk, in the region. This link between radicalism and perceived injustice was noted by the Supreme Court who criticised the Indian government's development policy as 'blinkered' (*The Times of India*, 21/07/2010) and argued that the inequality it was promoting exposed the country to the risk of terrorist groups, such as the Naxalites (ibid).

In pre-secession Telangana, the delegitimation of the Andhra Pradesh government was exacerbated by media reports of the killing of Naxalites by state security forces in response to the abductions, killings and acts of civil disobedience conducted by the Naxalites.[34] A survey undertaken by the *Times of India* (28/9/2010) on Naxalism illustrated the growing gulf between the perceptions of those in Telangana and politicians in the state and central governments. The survey, which involved 521 participants from five districts of Telangana, including Warangal, found that 58 per cent believed Naxalites were good for their area, and 57 per cent claimed that their killing by state security forces was unjustified.

The legitimation of democracy involved in the demand for secession sought to define the political means to bring about a more just and equal society. This involved discussions as to the meaning of the democratic ideal and how power should be exercised as a normative concern, as well as ideas on how to resolve problems with its exercise in practice. As Kannabiran et al. (2010: 79) note, the Telangana movement gave rise to 'many questions regarding participation, equity, power, inclusiveness, representation, democratic values, accountability, commitment [and] creative expression.' One protestor (as cited in Kannabiran et al. 2010: 79) argued: 'it [the Telangana movement] is a struggle against exploitation of a region. It is a struggle for justice. It is a struggle for adequate representation in social and political life.'

The protests for secession also contributed to a more nuanced political awareness in the region. Another participant in Kannabiran et al.'s (2010: 71) research observed: 'The political leadership is undemocratic and lacking in commitment without doubt but the mass base of the movement and the movement leadership has grown and developed on democratic values.' In seeking to secure legitimation for a separate state, therefore, the Telangana movement spread the discourse of democracy to broader sections of society. It also rendered the links between political engagement, the right to justification and the negotiation of risk more meaningful given that these aspects were seen as central to democratic legitimacy itself.

The influence of the norms associated with political protest in the Indian context was evident in the way in which the legitimation struggle associated with secession was undertaken. Telangana was not immune from the wider influence of Gandhi's advocacy of *ahimsa* (nonviolence) and *satyagraha* (literally, truth-force but, more commonly, passive resistance) in the struggle against the British. As Roosa (2001: 75) highlights, a delegation from the Congress Party travelled to Delhi during the Telangana struggle to request Gandhi's permission to adopt violent methods, along with the communists. Gandhi, openly weeping at times, had requested them to use nonviolence as best they could.

For Gandhi, acts of passive resistance included purificatory or penitential actions (prayers, pledges and fasts), acts of non-cooperation (boycotts, *bandhs* (strikes), fasts until death) and civil disobedience (defiance of specific laws) (Pantham 1987: 303). The influence of Gandhi's teachings on the methods adopted to highlight the legitimacy of the Telangana cause illustrates the way in which cultural norms and values impact upon the legitimation of democracy and the way democracy is practised, in particular contexts. As highlighted, these norms and values themselves emerge as a result of historic legitimation struggles.[35]

Protest methods adopted during the research period included regular *bandhs* which involved the closure of schools, shops and cinemas (*Times of India*, 8/1/2011; *Deccan Chronicle*, 22/2/2011), a rail *roko* (blockade) (*Deccan Chronicle*, 2/3/2011), student agitations (*Times of India*, 10/1/2011; 25/1/2011), *gherao* (encirclement) of a minister's home in Warangal (*Deccan Chronicle*, 9/1/2011) and relay hunger strikes (*Times of India*, 22/1/2011). The Chief Minister, Kiran Kumar Reddy, was also confronted by 2500 females in Warangal who vociferously shouted 'Jai Telangana' (*Times of India*, 11/2/2011).[36]

The idea of *tapas* or self-suffering which formed part of Gandhi's concept of *satyagraha* (Pantham 1987: 303) was also evident in the protests. Suicides for the 'T cause' were frequently reported in the media. These included deaths through self-immolation, particularly among students (*Times of India*, 12/2/2011), hanging (*Times of India*, 17/10/2010; 29/1/2011) and pesticide consumption (*Times of India*, 11/3/2011; *Deccan Chronicle*, 16/8/2010). A suicide note left by one of the deceased said that his suicide was meant to 'serve as a reminder…that people are willing to die for Telangana' (*Times of India*, 11/3/2011).

While the demand for a separate state sought recognition for the inter-regional inequality in pre-secession Andhra Pradesh, however, less attention was paid to the inequity within Telangana itself. As highlighted, access to land remains highly differentiated in Telangana. The Telangana armed struggle, land reform, reservations and the introduction of PRIs had shifted the balance of power in the villages somewhat; however, the enhanced ability of certain castes to corner opportunities for power associated with these initiatives meant that new dominant castes emerged to replace those which had previously prevailed.

It has been noted that, in Telangana, the emergence of the Forward Caste Reddys as a dominant force was significant. As Gundimeda (2009: 52) argues, '[i]t was by using both the land reform policies and the new administrative structures [PRIs] that the Reddys succeeded in consolidating their political power in [pre-secession] Andhra Pradesh.' Gundimeda (ibid.: 53) also notes that 'the *panchayat raj* system [in the pre-secession state] became a fresh avenue of power and prestige for the upper castes in general and Reddys, in particular.'

These shifts in power relations and access to land in the region allowed the Reddy caste to seize power from the Brahmins and the relatively powerful Backward Caste Gowdas (Gundimeda 2009: 52). The domination of political power and access to resources by the Reddys was strongly challenged by Scheduled Caste assertiveness (ibid.: 57).

This ongoing power struggle in Telangana contributed to concerns as to whether state secession would, in fact, lead to a more just access to resources, and greater recognition and representation for the risk exposure of the vulnerable, or would simply result in a shift in exploiters. This was asserted by Maringanti (2010: 37) who questioned: 'can capture of state power or redrawing of state boundaries by itself accomplish anything other than a change of actors – keeping the scripts intact?'

The pressure for secession was heightened in July, 2011 by the mass resignations of MLAs (Members of the Legislative Assembly) from the pre-secession Andhra Pradesh government.[37] This, along with the ongoing mass protests in the region, created the potential for a constitutional crisis in the pre-secession state unless the issue of secession was finally dealt with.

The secession was agreed to while the writing of my doctorate was in progress. This agreement was contrary to the recommendations of the Srikrishna Report which had reported its findings in January, 2011. The report (2010: 453) noted some justification for a separate state on the grounds of past injustices, particularly with regard to the failure to keep to the terms of the Gentlemen's Agreement. It also noted neglect in relation to water and education provision, as well as the unresolved issue concerning reservations for *mulkis* in public employment. Nonetheless, Srikrishna and his team (2010: 453) claimed '[s]eparation is recommended only...[if] it is unavoidable'. This was due to concerns for the potential turmoil that a divided state would bring for those from Coastal Andhra and Rayalaseema who had settled in Hyderabad and Telangana. Despite this, Telangana was announced as India's twenty-ninth state in June, 2014.

This chapter has explored the historical struggle for the legitimation of risk and democracy in the particular context of Telangana. It has examined the way in which this was continued through the contemporary demand for secession. The chapter also explored the particular context of risk within which Bt cotton was adopted in Warangal and Telangana. It argued that the protests against Bt technology and the demands for secession both represented attempts to gain recognition for the wider context of risk in the region and the inequality with which it is associated. The central concern of both mobilisations was the exercise of power with regard to risk negotiation and the practice of democracy. The methodology adopted to investigate the issue of Bt cotton within this context animated with demands for secession will now be explored.

Notes

1. All population figures for Andhra Pradesh are taken from the 2011 Census data. Available at: http://www.censusindia.gov.in/2011-prov-results/paper2/data_files/AP/4-fig-6.pdf. Accessed on 27/3/2017.
2. Available at: http://states-of-india.findthedata.org/q/1/4243/How-many-districts-are-there-in-Andhra-Pradesh. Accessed on 27/3/2017.
3. Available at: http://disastermanagement.ap.gov.in/historyofdisasters.aspx. Accessed on 27/3/2017.
4. Available at: http://www.telangana.gov.in/about/state-profile. Accessed on 27/3/2017.
5. The title of Nizam-ul-Mulk was granted to Chin Qulich Khan Asaf Jah and was retained by successive members of the Asaf Jahi dynasty which ruled Hyderabad.
6. Guntur district was occupied later in 1788 (Ramakrishna 1983: 2).
7. The Srikrishna Report (Srikrishna et al. 2010: 179) notes that British engineers, such as Sir Arthur Cotton, 'transformed lakhs of hectares [hundreds of thousands of acres] of barren lands in Coastal Andhra into a big rice bowl during the 19th century.'
8. The *diwani* system shared the essential features of a *ryotwari* system where revenue was paid directly to the government by cultivators (Vaikuntham 2004: 82).
9. Roosa (2001: 60) distinguishes three phases in the revolt: the first phase (1946–47) was an unarmed defensive struggle against large

landlords; the second (August 1947 to September 1948) occurred as a result of the Nizam's refusal to accede to the Indian Union. At this point, the insurgents took up weapons. The third phase began in September 1948 when the Indian army entered Hyderabad, overthrew the Nizam and launched a counterinsurgency against the communists. According to Roosa (ibid.), the struggle ended in 1951. Radical communism, however, remains a significant aspect of the legitimation struggle in contemporary Telangana.

10. The *Bhoodan* Movement or Land Gift Movement was a voluntary land reform movement in India. It was started by Acharya Vinoba Bhave in 1951 at Pochampally village in Telangana. The movement was promoted as an alternative to the violent land seizures associated with the communist attempt at securing greater equality in access to land. The idea soon spread throughout India. Although 5 million acres was pledged under the *Bhoodan* scheme nationwide, it is estimated that only 50 per cent of this land was redistributed. Available at: http://www.thehindu.com/news/national/bhoodan-land-at-centre-of-unparalleled-scam-says-ramesh/article3919255.ece. Accessed on 27/3/2017.

11. Report of the Commission on Farmers' Welfare, Government of Andhra Pradesh, p. 26. Available at: http://www.macroscan.org/pol/apr05/pdf/Full_Report_Commission_Farmer_AP.pdf. Accessed on 27/3/2017.

12. From *Reinventing Telangana: The Way Forward Socio Economic Outlook 2016*, p. 29. Available at: http://sakshieducation.com/Budget/TS/TS-Socio-Economic-2016.pdf. Accessed on 27/3/2017.

13. Figures cited here represent an average of the figures provided for North and South Telangana by these authors.

14. The merger was proposed primarily on linguistic grounds given that all three regions were Telugu-speaking because of their historic union within the Kakatiya Empire.

15. As highlighted in Chap. 3, Mech 12, Mech 162 and Mech 184 varieties were banned by the pre-secession state government in 2005 because of their widespread failure. Herring (2008: 154) argues that the successful demands for compensation in Warangal have led to ongoing fabricated (in his opinion) claims of crop failure given his view that compensation awards create 'incentives to claim poor results.' This book argues that the statement over-simplifies the wider context of risk, inequality and power struggle within which the conflict concerning Bt cotton is embedded in Warangal, as well as the highly variable performance of Bt cotton within this.

16. Naxalites take their name from an uprising of workers in the tea gardens of Naxalbari in Bengal in 1967 (Ray 1988: 82). The leadership coalesced in 1969 and split from the Communist Party India (Marxist) to form the Communist Party India (Marxist-Leninist) (Brass 1990: 299). As Ray (1988: 83) notes, however, the term Naxalite is commonly used to encompass all revolutionary communist groups.

17. Available at: http://www.telangana.gov.in/about/state-profile. Accessed on 27/3/2017.

18. Available at: http://www.census2011.co.in/census/city/398-warangal.html. Accessed on 27/3/2017.

19. Pre-secession, Andhra Pradesh was described as a 'leader in the MFI movement' (*Deccan Chronicle*, 14/10/2010). Commercialised MFIs were associated with a credit crisis in the state given the influence of the neoliberal rationale on these institutions (Shylendra 2006: 1961). This had seen the charging of interest rates of between 24 and 30 per cent (*Times of India*, 14/10/2010). MFIs have also been associated with unethical methods of debt recovery, such as the confiscation of title deeds and the use of intimidation (Shylendra 2006: 1959).

20. Alternative forms of irrigation involve rain-water harvesting through canals and tanks, in which the ground-water table is maintained.

21. This found that farmers within the category of 4–10 hectares (10–24 acres) earn an average monthly income of Rs 5479 per month, as opposed to consumption levels of Rs 4133 per month (Galab et al. 2009: 187). It should be noted, however, that farmers receive their income in three or four instalments at the end of the agricultural season as the crop is harvested and brought to market. This makes the management of their income particularly difficult. Similarly, the extent to which the consumption levels cited here include debt repayments is not clear. Fieldwork highlighted that income from the harvest is often first used to pay some of the more pressing debts.

22. The thesis upon which this book is based provides detailed data on leasing arrangements at the level of individual cultivators. This is available online at https://cora.ucc.ie/handle/10468/1688

23. Available at: http://www.census2011.co.in/census/district/126-warangal.html. Accessed on 27/3/2017. Pre-secession, Telangana had nine districts. This was increased to 31 in 2016. http://www.thehindu.com/news/national/telangana/Telangana-gets-21-new-districts/article15479100.ece. Accessed on 27/3/2017. The reorganisation aimed at

improving the effectiveness with which development and welfare programmes were implemented.

24. Warangal also includes thick black soils which retain moisture and are,
therefore, considered ideal for the cotton crop (Stone 2011: 390). In the
current study, cotton is cultivated on both red and black soils.

25. Available at: http://www.sourcewatch.org/index.php/Agriculture_in_
Warangal,_India. Accessed on 27/3/2017.

26. Paddy refers to rice still in the husk (Wade 1994: 40).
The delegitimation of Bt cotton in protests by cultivators relates to risks
associated with crop failure, pest attack and low yields. The impact
of these risks is, as the analysis will explore, highly differentiated as a
result of variations in access to resources. A sample of the grievances
raised in protests by Warangal cultivators is available at: http://www.
hindu.com/2004/10/16/stories/2004101604770500.htm; http://www.
indiatogether.org/2005/may/agr-apcotton.htm; http://yaleglobal.yale.
edu/content/bitter-sweet-gm-crops. Accessed 27/3/2017.

27. As highlighted previously, the region of Telangana is particularly susceptible to tropical storms, floods and droughts.

28. Full details are available at: http://csa-india.org/what-we-do/npm/.
Accessed on 27/3/2017. The main issues identified with NPM methods
are the inconsistent approach to the practice adopted by the government, inadequate numbers of training staff among implementing NGOs
and a lack of cooperation and organisation among farmers. Available at:
ftp://ftp.fao.org/sd/sda/sdar/sard/GP%20updates/pest_management_
India.pdf. Accessed on 29/3/2017.

29. This is most notably through the use of the leaves and oil of the neem
tree (a natural insecticide) and pheromone traps.

30. The power struggle at the meso-level of the state with regard to Bt cotton, and the policy decisions which are formulated as a result, are crucial to the way in which micro-level decisions on agricultural practice
are made. A meso-level analysis of the legitimation struggle concerning the Bt cotton debate is beyond the scope of this book but is available in the thesis on which this book is based at https://cora.ucc.ie/
handle/10468/1688

31. Brass (1990: 235) notes that the Telangana movement emerged in 1969
as a result of a court decision to invalidate the domicile rule which
reserved a number of places for Telangana *mulkis* in appointments to the
state electricity board.

32. A breakdown of the allocation of the 806 thousand million cubic (TMC) feet of water from the Krishna river by Melkote et al. (2010: 9) indicates that Telangana projects were allotted 266.83 TMC of water against a due share of 552 TMC. Mahbubnagar district in Telangana, an area known for its very high levels of agrarian distress, should have received 187 TMC of water but had received nothing as of 2010. Coastal Andhra, on the other hand, was alleged to have received several times more than its due share of 99 TMC.

33. Newspaper headlines in this regard include: 'Maoist terror resurfaces in T [Telangana]: Two Shot Dead, 3 Abducted by Naxals' (*Times of India*, 3/12/2010) and 'Naxal leader shot dead in cold blood' (*Times of India*, 28/1/2011). This was countered by reports of the abductions and violence of Naxalites: 'Naxals release three tribals' (*Times of India*, 7/12/2010) and 'Maoists torch 14 vehicles' (*Times of India*, 22/12/2010).

34. It is worth noting that 64 per cent of those questioned in the *Times of India* survey agreed with the Naxalites' methods of 'revolt against govt. apathy.' Despite the clear bias in the framing of the question, the finding suggests the growing legitimation of violence in the Telangana context. This was not in evidence during the struggle for secession which prioritised acts of civil disobedience and suicides over interpersonal acts of violence against opponents or representatives of state authority.

35. 'Jai Telangana' means 'Victory to' or 'Hail Telangana.' This was the slogan of the Telangana movement and was visible in graffiti throughout Hyderabad and the rural areas.

36. Details are available at http://www.oneindia.com/2011/07/04/telangana-resignations-72-mla-s-resign-more-to-join-aid0113.html. Accessed on 27/3/2017.

Bibliography

Ali, F., Panikker, K. M., & Kunzru, H. N. (1955). *Report of the states reorganisation commission*. Delhi: Government of India.

Bernstein, S. (2004). *The elusive basis of legitimacy in global governance: Three conceptions, Working paper: GHC 04/2*. Hamilton: Institute on Globalization and the Human Condition.

Brass, P. R. (1990). *The politics of India since independence*. New York: Cambridge University Press.

CESS. (2008). *Human development report 2007: Andhra Pradesh*. Hyderabad: Government of Andhra Pradesh.

Clark, I. (2007). Legitimacy in international or world society. In A. Hurrelmann, S. Schneider, & J. Steffek (Eds.), *Legitimacy in an age of global politics* (pp. 193–210). New York: Palgrave Macmillan.

Cohen, B. (2010). The historical context. In N. DeVotta (Ed.), *Understanding contemporary India*. London: Lynne Rienner Publishers.

Deshmukh, N. (2010). Cotton growers: Experience from Vidarbha. In R. S. Deshpande & S. Arora (Eds.), *Agrarian crisis and farmer suicides* (pp. 175–191). New Delhi: Sage.

Eyhorn, F. (2007). *Organic farming for sustainable livelihoods in developing countries? The case of cotton in India*. Zurich: vdf Hochschulverlag AG.

Fraser, N. (2008). *Scales of justice: Reimagining political space in a globalizing world*. Cambridge: Polity Press.

Galab, S., & Rao, N. C. (2003). Women's self-help groups, poverty alleviation and empowerment. *Economic and Political Weekly, 38*(12/13), 1274–1283.

Galab, S., Revathi, E., & Reddy, P. P. (2009). Farmers' suicides and unfolding agrarian crisis in Andhra Pradesh. In D. N. Reddy & S. Mishra (Eds.), *Agrarian crisis in India* (pp. 164–198). New Delhi: Oxford University Press.

Gaurav, S., & Mishra, S. (2012). *To Bt or not to Bt? Risk and uncertainty considerations in technology assessment*. Mumbai: Indira Gandhi Institute of Development Research.

Gundimeda, S. (2009). Dalits, Praja Rajyam party and caste politics in Andhra Pradesh. *Economic and Political Weekly, 44*(21), 50–58.

Gupta, S. P. (2002). *Report of the commission on India: Vision 2020*. New Delhi: Planning Commission, Government of India.

Haragopal, G. (2010). The Telangana people's movement: The unfolding political culture. *Economic and Political Weekly, 45*(42), 51–60.

Herring, R. J. (2008). Whose numbers count? Probing discrepant evidence on transgenic cotton in the Warangal district of India. *International Journal of Multiple Research Approaches, 2*(2), 145–159.

Inden, R. (1990). *Imagining India*. Oxford: Basil Blackwell.

Iyer, K. G., & Arora, S. (2010). Indebtedness and farmers' suicides. In R. S. Deshpande & S. Arora (Eds.), *Agrarian crisis and farmer suicides* (pp. 264–291). New Delhi: Sage.

Jangam, C. (2016). Dalit chronicles from the Telugu country. *Economic and Political Weekly, LI*(47), 25–29.

Kakar, S. (1995). *The colours of violence*. New Delhi: Penguin Books.

Kannabiran, K., Ramdas, S. R., Madhusudhan, N., Ashalatha, S., & Kumar, M. P. (2010). On the Telangana trail. *Economic and Political Weekly, XLV*(13), 69–81.

Kuruganti, K. (2006). Biosafety and beyond: GM crops in India. *Economic and Political Weekly, 41*(40), 4245–4247.

Maringanti, A. (2010). Telangana: Righting historical wrongs or getting the future right? *Economic & Political Weekly, XLV*(4), 33–38.

Mehta, J. L. (2005). *Advanced study in the history of modern India: 1707–1813*. New Delhi: New Dawn Press.

Melkote, R. S., Revathi, E., Lalita, K., Sajaya, K., & Suneetha, A. (2010). The movement for Telangana: Myth and reality. *Economic & Political Weekly, XLV*(2), 8–11.

Metcalf, B., & Metcalf, T. (2012). *A concise history of modern India*. New York: Cambridge University Press.

Michelutti, L. (2007). The vernacularization of democracy: Political participation and popular politics in North India. *Journal of the Royal Anthropological Institute, 13*, 639–656.

Misra, S. S. (2009). No pesticides, no debts. *Down to Earth, 400*, 26–28.

Mulligan, S. (2007). Legitimacy and the practice of political judgement. In A. Hurrelmann, S. Schneider, & J. Steffek (Eds.), *Legitimacy in an age of global politics* (pp. 75–89). New York: Palgrave Macmillan.

Omvedt, G. (1994). *Dalits and the democratic revolution: Dr. Ambedkar and the Dalit movement in colonial India*. London: Sage.

Pantham, T. (1987). Habermas' practical discourse and Gandhi's *Satyagraha*. In B. Parekh & T. Pantham (Eds.), *Political discourse: Explorations in Indian and western political thought* (pp. 292–310). London: Sage.

Qayum, A., & Sakkhari, K. (2005). *Bt cotton in Andhra Pradesh: A three-year assessment*. Hyderabad: Deccan Development Society.

Ramakrishna, V. (1983). *Social reform in Andhra (1848–1919)*. New Delhi: Vikas Publishing House.

Rao, V. M. (2009). Farmers' distress in a modernizing agriculture – The tragedy of the upwardly mobile: An overview. In D. N. Reddy & S. Mishra (Eds.), *Agrarian crisis in India* (pp. 109–125). New Delhi: Oxford University Press.

Rao, P. N., & Suri, K. C. (2006). Dimensions of agrarian distress in Andhra Pradesh. *Economic and Political Weekly, 41*(16), 1546–1552.

Ravi, T. (2015). Peasant struggles in Andhra Pradesh and Telangana: Reports from the field. *Review of Agrarian Studies, 5*(2), 110–125.

Ray, R. (1988). *The naxalites and their ideology*. Delhi: Oxford University Press.

Reddy, T. P. (2010). Distress and deceased in Andhra Pradesh. In R. S. Deshpande & S. Arora (Eds.), *Agrarian crisis and farmer suicides* (pp. 242–263). London: Sage.

Reddy, C. S., Jojaiah, K., Rao, N. V., & Narsaiah, I. (2012). Land and income inequalities in rural Andhra Pradesh. *The Marxist, XXVIII*, 50–74.

Renn, O. (2008). *Risk governance: Coping with uncertainty in a complex world*. London: Earthscan.

Revathi, E. (2009). 'Farmers' suicides in Andhra Pradesh: Issues and policy concerns. In S. M. Dev, C. Ravi, & M. Venkatanarayana (Eds.), *Human development in Andhra Pradesh: Experiences, issues and challenges*. Hyderabad: Centre of Economic and Social Studies.

Robinson, M. S. (1988). *Local politics: The law of the fishes: Development through political change in Medak district, Andhra Pradesh (South India)*. Delhi: Oxford University Press.

Roosa, J. (2001). Passive revolution meets peasant revolution: Indian nationalism and the Telangana revolt. *Journal of Peasant Studies, 28*(4), 57–94.

Shylendra, H. S. (2006). Microfinance institutions in Andhra Pradesh: Crisis and diagnosis. *Economic and Political Weekly, 41*(20), 1959–1963.

Sridhar, V. (2006). Why do farmers commit suicide? The case of Andhra Pradesh. *Economic and Political Weekly, 41*(16), 1559–1565.

Srikrishna, B. N., Duggal, V. K., Singh, R., Shariff, A., & Kaur, R. (2010). *Committee for consultations on the situation in Andhra Pradesh*. New Delhi: Government of India.

Srinivasulu, K. (2002). *Caste, class and social articulation in Andhra Pradesh: Mapping differential regional trajectories*. London: Overseas Development Institute.

Srinivasulu, K., & Sarangi, P. (1999). Political realignments in post-NTR Andhra Pradesh. *Economic and Political Weekly, 34*(34/35), 2449–2458.

Stone, G. D. (2011). Field *versus* farm in Warangal: Bt cotton, higher yields, and larger questions. *World Development, 39*(3), 387–398.

Sundarayya, P. (1972). *Telangana people's struggle and its lessons*. Calcutta: Desraj Chadha on behalf of the Communist Party (Marxist).

Tesoriero, F. (2005). Strengthening communities through women's self help groups in South India. *Community Development Journal, 41*(3), 321–333.

Vaikuntham, Y. (2004). *Studies in socio-economic and political history: Hyderabad state*. Hyderabad: Karshak Art Printers.

Vakulabharanam, V. (2004). Agricultural growth and irrigation in Telangana: A review of evidence. *Economic and Political Weekly, 39*(13), 1421–1426.

Venkateshwarlu, D., & Srinivas, K. (2000). *Debt and deep well: Status of small and marginal farmers in Warangal district.* Hyderabad: Care AP; ICNGO Programme AP; MARI.

Wade, R. (1994). *Village republics: Economic conditions for collective action in south India.* San Francisco: ICS Press.

Web-Sites

Rao, A. S. 95 per cent people below poverty line in Andhra Pradesh. *India Today*, 17/12/2011. http://indiatoday.intoday.in/story/95-per-cent-below-poverty-line-andhra-pradesh/1/164651.html. Accessed 26 Mar 2017.

5

Methodology: Legitimation and Ethics in Risk Research

In this chapter, the methodology adopted to undertake the study into Bt cotton in the three villages is explored. This discusses the way the villages were identified and participants selected. The adoption of a triangulated, mixed methods ethnographic approach to data gathering relates to attempts to derive information from as many perspectives as possible. This is given the controversy with which Bt technology is associated and recognition that perspectives on the theme are themselves mediated through power relations. The chapter also explores some of the very real risks involved in conducting research in a volatile and vulnerable context, such as Telangana.

The requirement for reflexivity on the positioning of the researcher within the research process is highlighted. This relates not only to the need for a careful negotiation of village power relations as part of the research process but also to the impact of the researcher's own 'positionality' (Pearson 2006: 308), and the power relation arising from this, upon the conduct of the research itself. Finally, the complex issue of the legitimation of research into risk and the particular ethical dilemmas associated with such work are explored.

© The Author(s) 2018   **133**
E.L. Desmond, *Legitimation in a World at Risk*,
https://doi.org/10.1007/978-981-10-6065-6_5

Locating the Research and the Villages

The three villages featured in this study were visited on alternate weekends throughout a 9-month period between June, 2010 and March, 2011. The timing of the fieldwork was chosen to coincide with the duration of the cotton season. The village research was conducted using a 'mixed methods' (Johnson et al. 2007: 112) approach which involved the collection of both qualitative and quantitative data. The latter related to a quantitative analysis of questionnaire responses, cost and income data related to the 2010/2011 season and levels of participant indebtedness.

Because of space constraints, and the problems of generalisation associated with the small scale of the study, the cost and income data of participants for the 2010/2011 season has not been included in this book.[1] However, some broad comments on the material outcome of the season for the participants involved are provided here.

As mentioned in the previous chapter, the 2010/2011 season in which this research was conducted was characterised by extensive flooding. The heavy rains led to an increase in weeds. There was also a high incidence of sucking pests which required pesticides. The floods also resulted in significant crop loss.[2] The poor prices being offered by traders because of the ostensibly poor quality of the cotton which was not destroyed led to general unrest among cotton cultivators,[3] as well as a spate of farmer suicides.[4] As highlighted in Chap. 4, these extremes of climate are by no means rare in Telangana and represent an important aspect of the agrarian crisis in the region.

The analysis highlighted that the costs of Bt cotton cultivation were higher than for organic and NPM cultivation. The comparison between the methods was explored for small and marginal cultivators given that NPM and organic cotton cultivators in the semi-medium and medium land-holding categories were unavailable in the villages studied. The average cost per acre for marginal Bt cotton cultivators in Bantala and Nandanapuram (N = 4) was Rs 15,000, while the cost per acre for the NPM cultivator in Nandanapuram (N = 1) was Rs 10,700 (due to the absence of the spend on pesticides). Organic cultivation for marginal cultivators was cheaper again (given the absence of spending on both pesticides and fertilisers) at Rs 9625 per acre (N = 4).

For small-holders, the costs per acre rose to Rs 21,625 for Bt cotton cultivators (N = 4), Rs 12,800 for the NPM cultivator (N = 1) and Rs 11,200 for the organic small-holders (N = 3). For most cultivators, cotton was the predominant crop, with a smaller proportion of paddy. For small and marginal cotton cultivators, paddy cultivation was for household use only.

The cost of labour was significant for all cultivators. The average labour cost per acre for Bt cotton cultivators in Bantala and Nandanapuram was Rs 9200 (N = 12). For the two NPM cultivators in Nandanapuram, the average labour cost per acre was Rs 8850 (N = 2). Given the more collective approach to cultivation in Orgampalle, however, average labour rates were the lowest of the study at Rs 7400. The reasons for the variances in labour rates between the villages will be explored in more detail in Chap. 6.

The study found that of the 12 Bt cotton cultivators, five made a loss, with a further Bt cotton cultivator just breaking even. Despite the catastrophic season, however, not one NPM (N = 2) or organic cultivator (N = 7) made a loss given the lower input costs involved.

The triangulated research methods adopted in this study involved the use of a number of different data sources in order to permit the corroboration of the data (Hammersley and Atkinson 1995: 214; Johnson et al. 2007: 113). These sources included semi-structured interviews in the villages over a 9-month period, questionnaires, field notes and observations, as well as interviews with academics, industry representatives, regulators, NGOs and politicians. Research was also undertaken in local libraries in Hyderabad and Warangal, and seminars in Indian Sociology were undertaken at Hyderabad Central University (HCU). Finally, a significant database of newspaper clippings was collected over the 9-month period.

This mixed methods, triangulated approach was adopted because of the recognition of the way in which the controversy surrounding Bt technology is mediated through power relations. The mixture of methods sought to mitigate the potential for the research to become directed by powerful interests through exploring the issue from a variety of perspectives. As Miller and Fox (2004: 36) note, 'looking at an object from more than one standpoint provides researchers…with more comprehensive knowledge about the object.' It was also felt that a sociological investigation which sought 'to consider multiple viewpoints, perspectives, positions, and standpoints' (Johnson et al. 2007: 113) was essential to understanding such a controversial topic. The mixed methods approach also reflected the complexity of the theme.

The qualitative interview data is supplemented by quantitative information. This includes the quantitative analysis of the results of a questionnaire conducted as part of the research interviews in order to obtain an overview of village perspectives. The results of this are included as Appendix 17. The questionnaire involved the use of a Likert Scale (with rankings ranging from strongly agree to strongly disagree) to assess the views of participants on statements related to Bt technology and democratic practice.

Again, due to space constraints, only selective findings from this questionnaire have been featured in the analysis. It should be noted that the five landless participants declined to respond to the questionnaires, claiming that they did not have the required knowledge. This highlighted the inappropriateness of the method for these participants given that the paperwork associated with the questionnaires was perceived as official and was immediately assessed, therefore, to be above their level of competence.

Quantitative data on accumulated debt levels is also included. This seeks to explore the often neglected aspect of the additional financial risk-taking which attempts to negotiate risk often entail. Given that indebtedness is an important factor in the risk of suicide, the data on participant indebtedness has been included here as the material basis for the study.[5]

The problems with generalisation arising from the small sample size of the study are recognised with regard to figures on debt. This book asserts, however, that micro-level studies of indebtedness allow the complexity and nuance of the issue of Bt cotton's contribution to the alleviation or exacerbation of cultivator risk to be more adequately explored than large-scale studies which assert the technology's effectiveness *overall*. This focus on the microlevel also permits an examination of the way in which indebtedness itself is differentiated and legitimated through power structures.

Information on the debt levels of participants was gathered as part of the interview process. It was noted that, while participants seemed comfortable in divulging information on their levels of debt, they were far more reticent about revealing specifics of the sources of the loans, often referring vaguely to 'members of their own caste' or 'others in the village.' Details of debt levels and the sources of borrowing for individual participants in each village are provided in Appendices 14, 15 and 16.

Two of the villages featured in the study, Bantala and Orgampalle, are located in south Warangal, approximately 10 kms apart, at a distance of 100 kms north of Hyderabad. Nandanapuram is situated in north Warangal, 110 kms from the other two villages and 210 kms to the north of Hyderabad. In order to preserve anonymity, village names are pseudonyms, as are the names of participants.

I first visited the villages of Bantala and Orgampalle during a 10-day pilot study undertaken in March, 2010. These trips were organised through a contact at the International Crops Research Institute for the Semi-Arid Tropics (ICRISAT) in Hyderabad and undertaken with a rural NGO, Crops Jangaon. The village of Bantala was not affiliated with an NGO for agricultural purposes. It had come to the attention of the NGO and ICRISAT as a result of reports of animal deaths by the villagers in 2008. Nandanapuram was first visited in July, 2010, in conjunction with the Deccan Development Society (DDS), an NGO based in Hyderabad. As highlighted, I had previously come across Nandanapuram in 2002 while conducting an Internet search on protests against GM crops during my employment with Syngenta, an agrochemical multinational, in Switzerland.

Despite my prior knowledge of Nandanapuram, however, village fieldwork initially focussed on Bantala. This was because I felt that the village of Nandanapuram was already a high-profile research site with strong links to the Deccan Development Society. The decision to focus on Bantala sought to overcome the potential for the research to be overtaken by the special interests of NGOs who might seek to assert a particular position through it.

In the case of the villages where NGOs were present, my concern was that NGO representatives or their employees in the villages would direct me towards participants who they knew had a favourable view of the NGO's approach, or that villagers would be guarded in expressing their own views. In an attempt to encourage candid responses, interviews were conducted with participants in their homes away from other villagers— often no easy feat given the tendency for groups to gather and attempt to participate in the discussion.

As highlighted, the Crops Jangaon representative who had initially accompanied me to Bantala had previously visited at the time of the animal deaths only, but was not directly affiliated with the village. I felt that the lack of prior publicity and the absence of an NGO affiliation with regard to cultivation in Bantala would enable me to gain fresh and, potentially, more authentic insights into a controversy which authors, such as Herring (2008: 155), argued was being instigated by NGOs for their own purposes.

After three months of just visiting Bantala, however, I decided to extend the breadth of the research to incorporate Orgampalle and Nandanapuram. This was due to a recognised lack of a basis for comparison of the perspectives of participants in Bantala. Also, given the legitimation of the technology in Bantala, a focus on this one village failed to represent the struggle with which I knew Bt cotton was associated in Warangal. Because Orgampalle is an organic village, it presented an opportunity to compare the experience of Bt cotton cultivators in Bantala with that of organic cultivators who delegitimate the technology. It also allowed the dimension of protests and the technology's delegitimation to be explored.

The decision to incorporate Nandanapuram was made not only because of my long-standing interest in the village but also because there were other aspects about the village which I considered important. Firstly, unlike Bantala and Orgampalle, Nandanapuram's population includes a small number of Forward Caste Brahmins. Because Brahmins represent the apex of the caste system, I was keen to understand the impact of their positioning within the village power structure. Secondly, Nandanapuram is located in north Warangal, and so is relatively distant from the other two villages. This meant that I could compare the performance of Bt cotton in two dispersed locations within the same district. Finally, Nandanapuram comprised both Non-Pesticide Management (NPM) and Bt cotton cultivators. This allowed me to compare the experiences of cultivators using two different cultivation methods within the same village.

Throughout the research period, half of each week was spent based at Hyderabad Central University (HCU). Classes from the Master's Sociology syllabus were attended between July and October, 2010. These included bi-weekly seminars on 'The Sociology of India' and 'Science, Culture and Society.' Private Telugu lessons were also undertaken at the university 3 times per week between July and September, 2010. These followed 3 months spent learning the language using CD tutorials prior to leaving for India. Due to the pressures associated with the research itself, however, the frequency of these language classes gradually decreased.

During my time in Hyderabad, literature reviews were conducted at the libraries of the International Crops Research Institute for the Semi-Arid Tropics (ICRISAT), the Center for Sustainable Agriculture (CSA), the National Institute for Rural Development (NIRD) and the Centre for Economic and Social Studies (CESS). Interviews were also conducted with academics, researchers, politicians, industry representatives, regulators and NGOs based in Hyderabad. These formed the basis of a wider meso-level analysis in my doctoral work.

Two conferences were attended in other states—an All-India Sociological Conference in Cuttack, Orissa (27–29/12/2010), and an international conference entitled 'Visions of Asia and the Challenges of Creative Social Theorizing' held in Bangalore, Karnataka (21–24/2/2011). I also presented a paper on my work at the Madras Institute of Development Studies in Chennai, Tamil Nadu (11/11/2010).

The second half of each week was spent conducting fieldwork in the three villages. A total of 26 participants were involved. These were selected with the help of 'key informants' (LeCompte and Schensul 1999: 86), an aspect of the research which will be covered in the next section. The starting point of the village analysis involved obtaining details from key informants on the land-holding and caste composition of the village (as provided in Appendices 4, 5 and 6). Key informants were then asked for suggested names of participants from each land-holding category (marginal, small, semi-medium and medium, as well as the landless), and caste category (Forward, Backward and Scheduled Castes)[6], as well as the names of female cultivators within these categories, wherever possible.

The aim of the selection process was to obtain a good cross-section of villagers within the dimensions of land-holding, caste and gender. The eight participants in both Bantala and Orgampalle are the result of this attempt. The ten participants in Nandanapuram reflect the larger size and greater diversity of the village. It was felt that this number of participants in each village was manageable given the timescale of the research and the desire to conduct a number of interviews with each participant throughout the season.

Most participants were interviewed up to three times throughout the 9-month period. In order to secure as broad an understanding as possible within the timescale, many informal unrecorded interviews were conducted with villagers, apart from those who officially participated. There were also a number of additional recorded interviews in Bantala. These included those with villagers who claimed their animals had died after grazing on Bt cotton (23/8/2010; 24/8/2010); the Agricultural Mediation Officer, a *panchayat* position (13/9/2010); and those who had received land during land reform (5/2/2011). Interviews were also held with Bantala's village vet (12/9/2010) and the Mandal Parishad Development Officer, as the person responsible for Bantala's development in the next tier of the PRI structure (4/10/2010). These have not been included as part of the formal analysis although they inform the findings.

Given my desire to avoid being positioned within village power structures, I was not based in the villages during the fieldwork. Instead, a basic hotel frequented by Indian NGO researchers, was identified. This was situated just outside Warangal city, equidistant between the villages in north and south Warangal.

Village interviews had to be conducted before 10 a.m. when participants began work in the fields. Given these early starts, I would arrive by train from Hyderabad the day before the 2-day village visits. The afternoon would be spent conducting library research at Kakatiya University in Warangal. Village field trips involved terrifying hour-long pre-dawn bus journeys in rickety vehicles driven at (their) top speed on uneven rural roads. Bus drivers would struggle to overtake other vehicles with seeming disregard for the oncoming lorries whose approaching headlights in the darkness seemed to signal the inevitability of collision. Somehow, we always managed to just make it back safely to our side of the road.

The bus journeys were followed by a relatively calm auto-rickshaw trip to the villages (or another bus trip in the case of Nandanapuram) as the sun rose. My translator and I would arrive as the cocks were crowing, and the females sleepily lighting fires and banging pots to prepare rice, *puri*[7] and *chai*[8] for breakfast. Interviews were conducted in the homes of participants while sitting cross-legged on the floor, often on an empty pesticide sack. When this became too uncomfortable, I would sit on a charpoy which, together with a TV, were often the main possessions evident in village homes.[9]

Most interviews lasted for up to an hour (including the time taken for translation). They were recorded using a digital recorder and transcribed in Hyderabad between village visits. The extracts included in this book are taken from over 60 village interviews, involving more than 500 pages of single-spaced, typed transcribed material.

Although onerous and time-consuming, transcribing between village visits allowed me to develop familiarity with the personal details of participants. It also meant that I could make notes of questions to be addressed which could be referred to prior to each field trip. Hearing the participant responses to my questions and the translations as part of the transcription work allowed me to become familiar with certain key Telugu words associated with the research. While I learned enough Telugu to be able to speak casually with villagers and to check certain aspects of the translation, my ability with the language was, unfortunately, not sufficiently developed to conduct the research interviews alone. Instead, I was obliged to locate translators.

During the village fieldwork, I worked with five different translators, all from Telangana, and all contacted through Hyderabad Central University. They included four males and one female who were aged between 20 and 30 years. Of the males, one was a Master's student in Anthropology, one a PhD student in Economics, one a Master's student in Sociology and one a Business graduate. The female was a PhD student in Hindi and a Hindi teacher. The female, the male PhD student and the Business graduate were all Backward Castes. The Anthropology and Sociology Master's students were a Scheduled Caste Mala and Madiga, respectively. Given my concern with anonymity, these translators were required to sign a confidentiality agreement (included as Appendix 3).

I soon realised that the caste of my translators was important given that, as will be discussed, the differences in their castes allowed me to observe the variety of interactions arising from the intersection of the caste of my translator with that of the research participants. This heightened my sensitivity to the social negotiation of caste in the Indian context. I would otherwise have been denied this insight given my limited opportunities to observe people, whose castes were known to me, interacting with each other. The change in translators which came about unintentionally for a variety of reasons, including illness, exam pressure and personal bereavement, also ensured that the potential for translators to bias the research noted by Mackenzie et al. (2007: 304) was minimised. It also served to strengthen my rapport with participants given that I was the constant presence throughout the 9 months.

The Negotiation of Power Relations in Social Research

The central objective of the research was to understand the way in which the adoption of Bt technology was legitimated and delegitimated by cultivators themselves as a concern for their risk negotiation and to analyse how this legitimation and delegitimation was mediated through village power relations. Within this, however, it was also recognised that ethnographic research itself involves a power relation which must be carefully negotiated and reflected upon as part of the research process.

Because Bt cotton is a controversial topic in Warangal, I was wary of being positioned in the debate in ways which would lead to bias in the responses I received from village participants. As Pearson (2006: 313) highlights, '[f]armers in the developing world are adept at modulating their responses according to who is doing the asking.' While I was introduced to the villages by NGOs, care was taken to highlight to participants that I had come to learn of the farmers' own perspectives on Bt cotton, both positive and negative, and that I had no personal affiliation with the NGOs or with any other interest groups. Translators were also informed that they should avoid phrasing questions in ways which might indicate any personal bias on the theme.

My removal from an Irish context where Bt technology is banned for commercial cultivation sought to secure a degree of distance from the values and norms associated with perspectives on Bt crops in Ireland in the absence of their cultivation. Locating the research in Warangal provided the opportunity to gain insight into Bt technology in a context where a Bt crop had been cultivated (officially) for almost a decade at the time of the research but where it was still a source of conflict. My approach sought to undercut the higher-level legitimation struggle in which the views of cultivators seemed to be marginalised and where an ongoing power struggle hindered attempts to establish the factual validity of information. To this end, I believed that my only alternative in researching the theme was to approach cultivators themselves.

The power of 'gatekeepers' who have the authority to 'grant or refuse access' to participants in ethnographic research is noted by Hammersley and Atkinson (1995: 64). Gatekeepers were a crucial aspect of the current study in that, not only did I need their permission to conduct the research in the villages, I also had to manage the potential for gatekeepers to intervene in the research in ways which would skew my findings. Consent to conduct the research was sought from the *sarpanch* (head person) of each village. In the case of Orgampalle, Crops Jangaon, the local NGO with whom the village is affiliated, served as an additional gatekeeper beyond the village, and permission had to be sought from them prior to each village visit. During my first research trip to Orgampalle, employees of Crops Jangaon had to be asked politely to leave interviews so that my translator and I could interact with village participants independently. They did not accompany us again. The consent forms used with regard to both the *sarpanch* and participants are included as Appendices 1 and 2, respectively.

As highlighted, throughout my time in the villages, I also worked with a number of 'key informants' (LeCompte and Schensul 1999: 86). Given that Orgampalle and Nandanapuram had been studied previously, information on the village composition relating to the numerical presence of particular castes in the villages, as well as their land-holdings, was readily available. In the case of Bantala, however, this had to be compiled with the help of the Village Secretary (a *panchayat* position) and the headmaster of the village school. This information for each village is provided in Appendices 4, 5 and 6.

This prework on the village composition was essential to ensuring a valid cross-section of participants along the dimensions of caste, gender and land-holding and to identifying their relative positions within the village power structure. In Orgampalle, key informants included a village elder (Pradnesh) and a Crops Jangaon employee who resided in the village (Prakash); in Nandanapuram, they involved an NGO employee (SEED) who resided in the village (Nand) and an assistant who did not participate in the research itself but greatly facilitated it.

Once the village composition was clear, key informants were asked for suggestions of participants of both genders, from specific caste and land-holding categories. This sought to ensure that I had a good cross-section of participants in terms of their relative power in the village and their likely exposure to risk. I also incorporated a number of participants from chance encounters in the fields to avoid the potential for key informants to exert too great a control over the selection process.

The characteristic segregation of castes into different wards in the Indian village, noted in Chap. 2, provides a material representation of village power relations. The more powerful castes and influential villagers are located in the centre of the village, along with the majority of the facilities. The *panchayat* (council) office and school are also centrally situated. The Scheduled Caste wards are positioned on the outskirts of the village and are noticeably less well maintained than the other wards. Village maps which provide the locations of individual participants within each village to indicate their relative positioning within the power structure are included as Appendices 8, 9 and 10.

The observation by Pemunta (2010: 4) that 'we can either disqualify our subjectivity as a hindrance to proper research, or turn it into our main research tool' is particularly insightful. I was aware of the impact of personal characteristics such as my gender, race, class and nationality (Pearson 2006: 308) on the research process. I was also sensitive to the way in which my personality intersected with the differing positions of participants within the village power structure (Wolf 1996: 17). Over time, reflexivity on this positionality itself was intentionally incorporated into the research process.

As highlighted, I avoided staying in the villages for fear of becoming positioned within the power structures in ways which would impact negatively upon the research. Nevertheless, after a number of months of visiting Bantala, I accepted an invitation to stay overnight with the family of a Scheduled Caste research participant (Ashna) to whom I had grown close. This was arranged for the *Bonalu* festival where I was granted the honour of bearing the *Dalit bonham* on my head to take to the Scheduled Caste shrine.[10] The decision to stay overnight was made not only due to my closeness to the family but also because I was curious to see what the reaction in the village would be. I also felt that, because it was just one night, any negative consequences could be redeemed.

News of my planned stay with the Scheduled Castes was greeted with shock in the other caste wards. This was articulated to me most directly by participants from the dominant and Forward Castes. One powerful Forward Caste participant (Pavan) argued that Scheduled Castes 'lived in slums' (field note extract, 30/8/2010) and that I should instead stay with them in the Forward Caste ward. Shortly after my stay, a dominant caste female, who had initially been a willing research participant, became impossible to locate for interviews and her previous contributions had to be removed from the analysis. I presume from the timing of her absence, and the new coldness in the response of her family to me, that her reluctance to continue was linked to my stay with Ashna, although this was never confirmed.

The reaction to my stay in Bantala reinforced my sense that the decision not to stay in the villages full-time had been the right one. While I was able to largely retrieve the situation, this would have been more difficult had I stayed in the villages on a regular basis. I was also aware that my positionality as a Westerner allowed me to secure a certain degree of tolerance in the overstepping of caste norms which I presume would not be granted to a local researcher. I took meals, for instance, with Brahmins and Madigas alike. This forbearance would, I believe, have been less forthcoming had I been a local researcher or permanently positioned within a particular caste ward.

Wolf (1996: 11) notes that positionality is multi-faceted and that different aspects offset each other. Being a lone female researcher was certainly a hindrance in certain aspects of the fieldwork, particularly with regard to finding suitable accommodation. It is also rare for females to travel alone in India which meant that I attracted considerable attention in rural areas. This was evident in the reporting of my visit to a cotton market in a local newspaper where I was described as an American scientist. Likewise, a newspaper reporter had to be politely requested to leave an interview which I had scheduled with the Mandal Parishad Development Officer (MPDO).[11] Other dimensions of my positionality, however, such as being middle-aged and 'foreign,' mitigated some of the obstacles presented by my gender.[12]

The heightened respect generally granted to Westerners in rural India should not be under-estimated when conducting research. It contributes to explanations of how the colonial enterprise was capable of securing a degree of legitimation within the Indian context. This respect appeared to be reinforced by my initially pale Irish skin and the association of fairness with Brahmanical purity.[13]

There was considerable deference paid to me on the basis of my skin colour and height in the rural areas. (At 5′9, I was significantly taller than the women and many of the men in the villages). This esteem was made explicit when an unknown elderly male approached me and touched my feet (a sign of respect) while I was waiting for an auto-rickshaw to take me to Bantala. One elderly lady (whose age was estimated to be over 100, although she was uncertain of the precise year of her birth) asked my translator in awe: 'are they all as tall and fair as her in her kingdom?' (field note extract: 20/6/2010). My appearance, at least initially, also carried with it a significant capacity to shock - children in Nandanapuram, quite disconcertingly, ran from me screaming when I first appeared in the village.

Once I became aware of this dimension of my positionality, I sought to reflexively monitor it in ways which would enhance the research. In the rural areas, this entailed always taking time to put people at their ease, as well as making efforts to assimilate through taking food with villagers, speaking in a low voice, wearing local dress, adopting the gestures used by those around me (including the surprisingly addictive waggle of my head to signify agreement), tying back and darkening my hair, tanning my skin and using my basic Telugu, particularly with children.

My positionality as a foreigner was, however, also used to gain access to high-profile participants, especially the politicians, industry representatives and regulators involved in the wider study. The ease of access associated with being a foreign female researcher in India is highlighted by the American anthropologist, Marguerite Robinson. Robinson (1988: 269) notes how she was welcome 'in both mud huts and marble mansions' during her fieldwork in pre-secession Andhra Pradesh. This, Robinson (ibid.: 269) argues, presents Western researchers in India with 'an unusual opportunity to listen, to learn, and to be heard in varied environments.'

Over time, my novelty value in the villages diminished significantly. This was facilitated by my gender which served to ease the potential power imbalance associated with being a foreigner in rural India. The view that females are seen as 'unthreatening or not official' in fieldwork is also noted by McDowell (1988: 85).

Relations with villagers were also eased through their often undisguised hilarity at my attempts to 'fit in.' They would look on in wry amusement at my gauche attempts to sit in lotus position and eat food with my fingers, or laugh delightedly at my speaking of Telugu with an Irish accent. Likewise, my own attempts to negotiate the risks of Indian rural life caused considerable amusement. In Bantala, there was much glee at my seemingly eccentric Irish habit of wearing wellington boots in conjunction with the elegant *salwar kameez*[14] to walk in the cotton fields. The boots were actually worn given my fear of the king cobra snakes in the village and my deluded notion that the sturdy boots would somehow offer protection against these.

As mentioned, the research also highlighted the need to take account of the impact of the positionality of translators upon the research process. In the current study, this was most evident in relation to caste. Mackenzie et al. (2007: 304) argue that translators can hinder 'mutual understanding' and 'potentially undermin[e] the validity of the research' (ibid.). Jacobsen and Landau (2003: 193) also note that there is a risk that the involvement of translators will lead to the 'transgressing [of] political, social or economic fault-lines of which the researcher may not be aware.'

Given the political unrest in Telangana, care was taken to ensure that my translators were from the Telangana region. Despite the fact that my inability to speak Telugu fluently was frustrating because it meant that my rapport with participants was mediated, working with translators heightened my learning in other ways. This was particularly with regard to the nuanced insight which the translator's interaction with participants permitted in relation to caste. Without a translator, I would have struggled to gain such an awareness given the strict segregation of castes in the villages.

I noticed, for instance, that the Scheduled Caste Madiga and Mala translators referred to Forward Caste Brahmin and Reddy participants in the polite form of Telugu, while other participants would be addressed in the more familiar form.[15] There was also a marked reluctance by the Mala translator to interrupt Brahmin and Reddy participants in order to provide the translation for fear he would be 'scolded' (field note extract, 15/8/2010).

The hierarchy between the Scheduled Castes themselves was highlighted by the Mala translator's refusal to eat a meal at the home of the Madiga participant (Ashna) where I was having lunch. This was because, he argued, 'his mother would scold him' (field note extract, 24/8/2010). Further clarification established that this refusal to eat with the Madiga family was due to the perceived potential for pollution through sharing food with Madigas.

The practice of Sanskritisation identified by Srinivas (1966: 6) was also evident in the Madiga translator's claim to have adopted Sanskrit words in order to fool a Brahmin participant (Charan) into believing that he was a higher caste (field note extract, 22/11/2010). This, I suspected, was also a reflection of his discomfort within the Brahmin's home given that, as a Scheduled Caste, his presence there would have been prohibited.

This sense was reinforced when the female Brahmin participant (Rashi) insisted her interview be conducted outside when I arrived with the same Madiga translator (field note extract, 21/11/2010). She also made frequent reference to her need to bathe (her contact with my translator would have been regarded as 'polluting') and seemed uncomfortable throughout the interview—until the issue of Telangana was raised. This led to an animated discussion between them both which I was obliged to interrupt in order to obtain a translation. The exchange highlighted the way in which the demand for secession transcended the power relations within Telangana itself.

The Legitimation of Research into Risk

The ethical dilemmas created by research into risk raise complex questions concerning the legitimation of research itself. It is acknowledged that ethical principles of 'beneficence, integrity, respect for persons, autonomy and justice' (Mackenzie et al. 2007: 299) represent the cornerstone of legitimate research practice. It is also recognised however that, as Mackenzie et al. (ibid.: 300) note, 'the articulation of these principles in ethics guidelines is often highly abstract.' This means that researchers are constantly obliged to reflexively legitimate their own research and their approach to it, both in the field and away from it, as part of the research process. This is particularly true of research into risk, given that it explores ambiguity and involves participants who are vulnerable. Here, the distinction between 'procedural ethics[16] and "ethics in practice"' noted by Guillemin and Gillam (2004: 261) is especially significant.

Ethical issues are relevant in risk research from the earliest stage of obtaining informed consent from vulnerable participants (Mackenzie et al. 2007: 299). This is particularly a concern given the ambiguity of risk and the uncertainty related to the consequences of the research. Because of this, pseudonyms have been used for villages and participants to try to preserve anonymity as far as possible. This is despite the insistence of village participants and the *sarpanches* that actual names could be used. The use of real names has been resisted, however, in order to eliminate the (unlikely) possibility for any unintentional harm to come to participants as a result of the research. Pseudonyms have also been adopted given that the debt levels, and borrowing sources, of participants are discussed. I felt participants were likely to be more honest in this regard if they knew their real names would not be revealed.

Research into risk generally involves fieldwork in locations characterised by their poverty given that these are the locations associated with the heightened risk exposure of inhabitants. In terms of the current study, the objectives of the research were translated into Telugu for all participants at the beginning of the research, as was the consent form (Appendix 2). Participants were then asked to sign or, in the case of illiteracy, to place a thumb print on the form. The acceptance of a thumb print as an indication of consent is, in itself, ethically questionable given the potential for exploitation which it highlights. In the case of risk research, however, it is the input of these most vulnerable of participants which is of the most vital significance. Hence, the ethical concerns regarding the inclusion of illiterate participants in research have to be balanced against the very questionable ethics entailed in their exclusion from it.

The awareness of the differentiation of risk exposure between the researcher and research participants is also ethically problematic. The concern raised by feminist researchers that research itself is exploitative

(Chase 1996: 49) is particularly heightened in research into risk. Within this, the complex issue of reciprocity and how participants should be compensated for their involvement (Skeggs 1994: 81; Wolf 1996: 19; Pittaway et al. 2010: 238) is especially acute. This ethical concern is exacerbated by the fact that the researcher is often in a position to significantly alleviate the risk exposure of some of the research participants, at least in the short term.

The awarding of cash gifts for interviews may serve to somewhat ease a researcher's guilt and ethical discomfort; it may also, however, negatively impact upon the research, reinforcing a power relation and potentially encouraging participants to respond in ways which it is felt will meet the researcher's approval. This is highlighted by Bolognani (2007: 283) who argued that '[m]onetary reward, which has become more popular recently in social research…could have secured more interviews but not necessarily honest accounts.' This is especially true when research involves vulnerable participants in contexts marked by their poverty. There is also the ethical dilemma involved in choosing who to help, and the realisation that your assistance, such as it is, will not contribute to a longer-term solution in the way that your research, if properly conducted, might.

I refrained from offering financial recompense for interviews, despite the queries of some of the landless participants, in particular, as to whether they would be paid. I did, however, find other less obvious ways to provide assistance which could not be directly linked to the research process. This involved sharing auto-rickshaws with landless villagers and passing them significantly more than my share of the fare. Gifts of cash and *sari* material were also subtly passed to the more vulnerable on my final visit.

Leaving the villages at the end of the research is also difficult. This relates to, as Wolf (1996: 10) notes, the privileged position of the researcher who can simply leave the site. This is more directly evident in research into risk where the researcher is leaving a context where some research participants are at risk due to their poverty. Attachments are formed during the research process, and leaving is particularly hard given the language barrier and geographical distance involved which makes remaining in contact impossible.

At the end of the research, a leaving ceremony was held in each village. This involved a village gathering and the presentation of a large cake and 'Certificate of Thanks' which was translated into Telugu and which featured photographs specific to each village. A thank you speech was made, and simultaneously translated, where the objectives of the research were again clarified, and humorous stories relating to my time in each village were exchanged.

Ethical dilemmas do not end, however, when the researcher leaves the field. Concerns for exploitation also relate to issues of representation and authorship (Patai 1991: 139; Wolf 1996: 33–34; Bolognani 2007: 288). This is particularly true when participants are geographically distant from the researcher, and where prior representation has been limited. To this end, a summary of the main points from the research findings, translated into Telugu by a former translator, has been posted to the key informants in all three villages for onward dissemination, along with photographs from my time there.

There are clear issues of representation in the use of translators given that the words of cultivators are not directly used. Mackenzie et al. (2007: 304) note that '[p]oor translation can hamper the kind of mutual understanding required for ethical research.' The use of a mixed methods approach to corroborate translated material and the enlistment of my own limited Telugu to check key aspects of the translation sought to address this concern in the current research. The fact that a number of translators were used, though not a deliberate strategy, was also no doubt helpful in limiting the impact of poor translation or translator bias. It was also noted that representation is a concern even when researchers and participants speak the same language. Additionally, as highlighted in this particular study, the use of translators revealed a dimension of the research into village power relations which would not otherwise have emerged.

The concern for representation also relates to the need to ensure that all positions are fairly presented, even if they are not positions which researchers themselves would readily legitimate. Beetham (2013: 11) argues that a focus on legitimation as a framework for analysis means that the exploration must involve 'analysing people's beliefs and examining their reasons for holding them.' This seeks to understand 'legitimacy-in-context' (ibid.: 14) as the conventions and beliefs which are being drawn upon from within a particular society to legitimate power relations and practices, rather than assertions of the researcher's own beliefs.

The focus on legitimation engages researchers in ensuring that all perspectives are themselves legitimately represented and then using this overview as the basis for a broader understanding of what the struggle between competing perspectives represents within that context. This is not value-free given that knowledge construction never is; however, it does mean that the researcher's values are reflexively monitored as far as possible and that attempts are made to challenge them through the design of the research process itself.

As Scheper-Hughes (1995: 147) notes, research is 'necessarily flawed' given the researcher's imperfect self-knowledge and ability to fully transcend his or her positionality and values; however, Scheper-Hughes (ibid.) also argues that it is 'good enough' if it seeks to push the boundaries of the researcher's self-awareness further and leads to new insights, not only on the theme of the research but also on the process of knowledge construction in research itself.

As part of the concern with ethics in the construction of knowledge pertaining to the 'epistemic gap' (Desmond 2014: 13) which risk represents, it is important that the research process is designed in ways which ensure that researchers are confronted with their own ideological blindspots and assumptions as far as possible, and challenged to alter them, rather than simply reproducing them. In this way, researchers learn from the experience of seeking to understand the reasoning of others in a different culture and, in doing so, transform their own self-understanding. Legitimate, ethical research should, therefore, represent a 'genuine internal revolution' (Baszanger and Dodier 2004: 14) for the researcher.

The effort to deliberately challenge my assumptions and the norms of my own cultural conditioning in the current research is reflected in the attempt to locate it in a context such as India. Nothing can prepare a neophyte researcher for fieldwork in India. Field trips involved extensive travel and hours spent at crowded railway stations awaiting delayed trains. There was also the heat and relentless questioning by strangers ('where are you from?'; 'why are you here?').

The Telangana rural environment, and its poverty, also involved aspects which added to the discomfort of fieldwork. These included risks related to scorpions, mosquitoes and king cobras, as well as the floods which, at times, made travel to the villages difficult. There was also my inability to drink village water (and my initial reluctance to eat village food due to health concerns), as well as the limited availability of hygienic toilet facilities. The political volatility contributed to the constant fear of Naxalite attacks on train journeys, and field trips had to be organised around the frequent strikes (*bandhs*) and road *rokos* (blockades). Fieldwork was often a lonely experience given that travel and hotel stays were undertaken in the absence of a translator and involved trying to make myself understood in my limited Telugu.

Despite these difficulties, however, the experience in the villages profoundly transformed my understanding of the world in ways which I am still attempting to articulate through my work. The beauty of the Indian country-side, the hospitality of the villagers, their sense of mischief and their wisdom, as well as the challenge to my own identity which adjustment to such a radically new, chaotic and diverse culture entailed, meant that the discomfort of field trips was more than compensated for by the richness of the fieldwork experience.

It was in the trips to the villages that a greater appreciation emerged of the way in which India's ancient philosophical and imperial heritage continues to inform the contemporary struggle of cultivators to establish a legitimate exercise of power in their negotiation of risk. It was also, paradoxically, in the micro location of the villages that a strong sense began to form of the way in which the conflict concerning Bt cotton in Warangal is a microcosm of a wider global struggle for legitimation with regard to the negotiation of risk more generally.

Through this research, I came to understand that my struggle to resolve my personal uncertainty with regard to Bt technology represents my own attempt to contribute to the legitimation process involved in negotiating a world at risk. While my disquiet regarding Bt technology arose as a result of my employment with Syngenta, my attempts to resolve it have led to a much more nuanced understanding of global risk more generally, as well as the complex power relations which must be considered as part of attempts to construct knowledge in relation to such risk.

This changed understanding has heightened my awareness of the interconnectedness of risk society and the urgency with which we now seek to learn from each other in order to resolve the 'non-knowledge' (Beck 2009: 115) associated with risk. These efforts to understand are contributing to a greater critical consciousness of power at local, national and global levels and raising questions of legitimacy as to how it is being exercised in response to risk. The outcomes of these evaluations relating to the exercise of power will themselves determine humanity's future survival in a world at risk, not only as an existential but as a moral concern.

In his ethnographic study in Ethiopia, Geleta (2013: 9) notes that he 'was not only interviewing [his] informants, but they were actively engaged in interviewing [him].' In the case of studies into risk, this can present particular ethical issues. This was evident in the questions from Bantala cultivators regarding the experience of those adopting Bt technology in Ireland. Without wishing to add to the anxiety which these participants were already expressing, I was obliged to respect their autonomy and to respond honestly that the technology was banned in Ireland due to concerns for its safety. This information was often met with thoughtful silence.

It would be impossible for me to predict the ways in which my encounter with participants impacted upon their perspectives and their future risk negotiation strategies any more than they could be sure of how they individually informed the conclusions which I reached as a result of our interactions. Cultivators in Nandanapuram were, however, notably moved to learn that it was Internet reports of protests in their village which had come to my attention many years earlier in Switzerland which had inspired the current study. The situation highlights the incredible inter-connectivity of risk society and the powerful force for collective learning, as well as risk, which it represents.

This chapter has presented the methods used in order to obtain the data involved in the analysis. The use of a triangulated, mixed methods approach was adopted in order to gain as broad a perspective as possible into what is recognised as a controversial theme. The importance of reflexivity on the positionality, not only of researchers themselves but also of translators, was discussed. It was argued that such reflexivity can provide valuable insights which enhance the research. Finally, the significant issue of ethics with regard to research into risk is explored. This highlights the importance of research which seeks not only to legitimately represent participant perspectives as far as possible but also to challenge the researcher's own values, assumptions and self-identity. The chapter argued that knowledge construction associated with ethical research is vital if we are to learn from each other in ways which support the negotiation of risk as a local, national and global concern. This learning, and the critical awareness of the exercise of power which it entails as a concern for its legitimacy, is recognised as crucial to establishing a normative basis for the negotiation of global risk as a collective of humanity. In the next chapter, the first of the two analysis chapters, the legitimation of risk in the villages will be explored.

Notes

1. They are, however, available in Chap. 6 of the doctoral thesis from which the research for this book is derived, available at: https://cora.ucc.ie/handle/10468/1688/
2. The *Deccan Chronicle* (9/12/2010) reported that around 25,000 hectares (60,000 acres) of cotton were damaged due to flooding. Cultivators claimed that 30 per cent of the crop had been lost (*Times of India*, 7/1/2011).
3. Cultivators in Warangal were involved in an altercation with cotton traders because of the low prices being offered for the crop (*Times of India*, 30/10/2010).
4. Following the flooding, farmer suicides were reported frequently in the media (*Deccan Chronicle*, 18/12/2010; 26/12/2010; 27/12/2010; 5/1/2011; *Times of India*, 8/10/2010; 9/12/2010; 10/12/2010; 18/12/10; 20/12/2010; 30/12/2010).
5. It is recognised that the causes of farmer suicides encompass a complex intersection of a number of structural issues, such as the integration of Indian agriculture into the world market, the deceleration of agricultural growth, the frequency of crop loss, the production of commercial crops and the decline in commodity prices, to name but a few. There are also non-economic factors, such as shame and embarrassment (Vasavi 2012: 22). It is acknowledged, too, that suicides are 'just a symptom [of the agrarian crisis] and not the disease' (Sainath 2007) and that, as Vasavi (2012: 22–23) highlights, neither Bt cotton nor indebtedness are sufficient to explain their causes. The current book, however, is a study of risk with specific reference to Bt cotton, not of suicide per se. As such,

it seeks to explore Bt cotton's contribution to alleviating or exacerbating indebtedness, particularly among small and marginal farmers, given that these categories of land-holder have been found to be most at risk of suicide and that indebtedness has been found to be an important factor in relation to such risk. The book does not seek to assert a straight-forward relationship between Bt cotton, indebtedness and farmer suicides; it is simply looking at this particular aspect as part of the Bt cotton debate while acknowledging the much broader picture which the issue of farmer suicides encompasses.

6. Participants from among Scheduled Tribes and Muslims were, unfortunately, not sought out given my own lack of knowledge regarding the significance of these groups in terms of their risk exposure in Telangana at the time of conducting the fieldwork. For the sake of simplicity, the category of Backward Caste has also not been sub-divided into Other Backward Castes (OBCs) as the more marginalised and deprived groups within the heterogenous Backward Caste category. Instead, the ranking of Backward Castes in the villages has been conducted according to the numerical presence and land-holding within the particular village. These aspects, as Frankel (2005: 6) also notes, help to define the local power structure given that the land-holding and numerical presence of castes in the villages coincides with their dominance (refer to Appendices 4, 5 and 6).

7. *Puri* is an unleavened, deep-fried Indian bread.

8. *Chai* is a spiced Indian tea. In the rural areas, this was served in scalding stainless steel tumblers.

9. A *charpoy* is an Indian bed consisting of a frame strung with woven tapes.

10. *Bonalu* is one of many village festivals in Warangal. It involves making offerings to the goddess *Kali* to ward off disease. In Bantala, hundreds of chickens were sacrificed to the goddess and their claws hung from trees at makeshift shrines. Brightly decorated earthenware pots bearing candles were carried on the heads of females to the shrines, while males, excitedly banging drums, accompanied them. The pots of each caste grouping bore distinctive markings, and castes in Bantala had separate shrines and processions conducted at different times throughout the day within their own caste wards.

11. The office of the Mandal Parishad had contacted the local newspaper to inform them that this interview was to take place. The reporter was waiting when I arrived for the interview with my translator. He was asked to leave given my concern that his presence would skew the interview or

allow the MPDO to simply use it as a public relations exercise. I also feared that my presence would become higher profile locally than I wished it to given my concerns for village anonymity and the impact of such newspaper coverage on my ability to build trust and rapport with villagers.

12. Because I was foreign and older, the fact that I was travelling alone seemed to be tolerated by men, and I never felt threatened. This was contrary to the views expressed to me by a number of Indian women who told me that they would not feel safe travelling on their own given the challenge this would represent to their cultural conventions.

13. As Bayly (1999: 173–175) highlights, there is a strong racial dimension to caste. The white-skinned Aryans who migrated from Central Asia circa 1500 B.C. were depicted as the 'pure' race, as opposed to the darker-skinned indigenous Dravidians. A fair complexion was subsequently asserted by high castes to support their claim to superiority on the basis of their purportedly 'pure' Aryan blood.

14. A form of Indian dress which involves loose trousers, a long shirt and a scarf (*chunni*).

15. Like many European languages, Telugu has a polite form of 'you' which is used for the elderly or those deemed socially superior. In Telugu, '*meeru*' is the polite form, while '*nuuvu*' is used for those deemed to be at an equal or socially inferior level.

16. Procedural ethics relates to the ethical requirements of funding bodies or academic institutions. In the case of the current research, this involved seeking ethical approval for the project from the Social Research Ethics Committee as part of the ethics procedures of University College Cork and as a requirement for the funding from the Irish Research Council.

Bibliography

Baszanger, I., & Dodier, N. (2004). Ethnography: Relating the part to the whole. In D. Silverman (Ed.), *Qualitative research: Theory, method and practice* (pp. 9–35). London: Sage.

Bayly, S. (1999). *Caste, society and politics in India from the eighteenth century to the modern age*. Cambridge: Cambridge University Press.

Beck, U. (2009). *World at risk*. Cambridge: Polity Press.

Beetham, D. (2013). *The legitimation of power*. London: Palgrave Macmillan.

Bolognani, M. (2007). Islam, ethnography and politics: Methodological issues in researching amongst West Yorkshire Pakistanis in 2005. *Social Research Methodology, 10*(4), 279–293.

Chase, S. E. (1996). Personal vulnerability and interpretative authority in narrative research. In R. Josselson (Ed.), *Ethics and process in the narrative study of lives* (pp. 45–59). London: Sage.

Desmond, E. (2014). *The legitimation of risk and democracy: A case study of Bt cotton in Andhra Pradesh, India.* Cork: University College Cork. Available at: https://cora.ucc.ie/handle/10468/1688/

Frankel, F. R. (2005). *India's political economy 1947–2004.* New Delhi: Oxford University Press.

Geleta, E. (2013). The politics of identity and methodology in African development ethnography. *Qualitative Research, 0*(0), 1–16.

Guillemin, M., & Gillam, L. (2004). Ethics, reflexivity, and "ethically important moments" in research. *Qualitative Inquiry, 10,* 261–280.

Hammersley, M., & Atkinson, P. (1995). *Ethnography: Principles in practice.* New York: Routledge.

Herring, R. J. (2008). Whose numbers count? Probing discrepant evidence on transgenic cotton in the Warangal district of India. *International Journal of Multiple Research Approaches, 2*(2), 145–159.

Jacobsen, K., & Landau, L. (2003). The dual imperative in refugee research: Some methodological and ethical considerations in social science research on forced migration. *Disasters, 27*(3), 185–206.

Johnson, R. B., Onwuegbuzie, A. J., & Turner, L. A. (2007). Toward a definition of mixed methods research. *Journal of Mixed Methods Research, 1*(2), 112–133.

LeCompte, M. D., & Schensul, J. J. (1999). *Book 1 – Ethnographer's toolkit: Designing and conducting ethnographic research.* London: Altamira Press.

Mackenzie, C., McDowell, C., & Pittaway, E. (2007). Beyond 'do no harm': The challenge of constructing ethical relationships in refugee research. *Journal of Refugee Studies, 20*(2), 299–319.

McDowell, L. (1988). Coming in from the dark: Feminist research in geography. In J. Eyles (Ed.), *Research in human geography* (pp. 154–173). Oxford: Blackwell.

Miller, G., & Fox, K. J. (2004). Building bridges: The possibility of analytic dialogue between ethnography, conversation analysis and Foucault. In D. Silverman (Ed.), *Qualitative research: Theory, method and practice* (pp. 35–55). London: Sage.

Patai, D. (1991). U.S. academics and third world women: Is ethical research possible? In S. B. Gluck & D. Patai (Eds.), *Women's words: The feminist practice of oral history*. London: Routledge.

Pearson, M. (2006). 'Science,' representation and resistance: The Bt cotton debate in Andhra Pradesh, India. *The Geographical Journal, 172*(4), 306–317.

Pemunta, N. V. (2010). Intersubjectivity and power in ethnographic research. *Qualitative Research Journal, 10*(2), 3–19.

Robinson, M. S. (1988). *Local politics: The law of the fishes: Development through political change in Medak district, Andhra Pradesh (South India)*. Delhi: Oxford University Press.

Sainath, P. (2007). *The farm crisis: Why have over one lakh farmers killed themselves in the past decade?* New Delhi: Speaker's Lecture Series.

Scheper-Hughes, N. (1995). The primacy of the ethical: Propositions for a militant anthropology. *Current Anthropology, 36*(3), 409–440.

Srinivas, M. N. (1966). *Social change in modern India*. New Delhi: Orient Longman.

Vasavi, A. R. (2012). *Shadow space: Suicides and the predicament of rural India*. New Delhi: Three Essays Collective.

Wolf, D. L. (1996). Situating feminist dilemmas in fieldwork. In D. L. Wolf (Ed.), *Feminist dilemmas in fieldwork*. Oxford: Westview Press.

6

Analysis I: Risk, Power and Bt Cotton in the Villages

This chapter explores the legitimation of risk in the villages. This is undertaken through examining the reasons provided by cultivators for their adoption or rejection of Bt cotton as a means to negotiating the wider risk of the agrarian crisis in Warangal. The analysis examines the way in which the legitimation and delegitmation of Bt technology is mediated through village power relations, as defined by land-holding, caste and gender. The exploration of power relations adopts Henrich's (2001: 997) conformist and prestige biases which were first applied by Stone (2007: 71) to Bt cotton in Warangal.

The chapter also investigates the relative indebtedness of participants and the intersection of this with village power relations. As well as indicating the risk exposure of participants, the analysis of accumulated debt levels over time gives some idea of the relative risk exposure of Bt cotton cultivators when compared to those adopting the alternative methods of organic and Non-Pesticide Management (NPM) cultivation.

© The Author(s) 2018
E.L. Desmond, *Legitimation in a World at Risk*,
https://doi.org/10.1007/978-981-10-6065-6_6

The Villages: Power Structures and Risk Profiles

Orgampalle and Bantala are both relatively isolated, located off a main road at the end of an unpaved track entailing a 15-minute auto-rickshaw journey; Nandanapuram is more accessible in terms of infrastructure and was reached following a one-hour bus journey, then a second half-hour bus trip from a nearby town. Given that the second bus was often up to an hour late, however, Nandanapuram at times proved more difficult than the others to reach.

Village maps have been included as Appendices 8, 9 and 10. All three villages have a paved main street at their centre. In Orgampalle, all houses are located along this main street. The villages of Bantala and Nandanapuram are bigger and are divided into caste wards. In the Scheduled Caste wards, the streets are not paved; instead, houses line a dirt track which becomes muddy in the rains. All three villages are electrified.

In Bantala and Nandanapuram, the *panchayat* office is located in the centre of the village, as is the school. Given the small size of Orgampalle, the *panchayat* office is located in a neighbouring village. A number of temples, whose walls are painted with vividly coloured depictions of Durga, Ganesh and, in the case of Bantala, a large king cobra, are dotted throughout the villages.[1] There is strict segregation in worship in Bantala and Nandanapuram, and Forward, Backward and Scheduled Castes attend separate temples. As the previous chapter highlighted, castes also celebrate festivals, such as *Bonalu*, within their own caste wards. There is evidence of Indiramma houses in various styles and stages of completion throughout all of the villages. The cultivated land of the inhabitants surrounds the villages.

On the outskirts of Bantala, a large canal has been excavated. The tributaries to carry the water to the cultivated land have not been completed, however, due to the absence of funding, and the canal remains unused. There is a water purification plant in the village for drinking water and a number of water tanks in the caste wards. In Nandanapuram, borewells are visible on the village lands suggesting their more general dispersal than in Bantala.

Village Risk Profile and Power Structure

Table 6.1 provides a brief overview of some of the key characteristics of the villages with regard to their composition and the context within which the cultivation of cotton is undertaken.

From Table 6.1, it can be seen (and was noted by villagers themselves) that Backward Castes are dominant in both Bantala and Orgampalle (BC Kuruma and Mudhiraj, respectively). The absence of a dominant caste in the case of Nandanapuram was asserted by the key informants and the research participants.

Table 6.1 The villages at a glance

Village	Bantala	Nandanapuram	Orgampalle
Population	2800 people (428 households)	3500 (971 households)	202 (52 households)
Location	South Warangal (10 kms from Orgampalle)	North Warangal (110 kms from Bantala and Orgampalle)	South Warangal (10 kms from Bantala)
Caste composition	FC—1 per cent; BC—59 per cent; SC—40 per cent	FC—1 per cent BC—59 per cent SC—40 per cent	FC—0 BC—100 per cent SC—0
Dominant caste	BC Kuruma (51 per cent of land holding)	No one caste is dominant. Majority of village land held by SC Madigas (37 per cent) and BC Gowdas (26 per cent)	BC Mudhiraj (82 per cent of land holding)
Sarpanch (head-person)	BC Kuruma female	BC Gowda female (previously SC Madiga male)	BC Mannuru Kapu male (in neighbouring village)
Percentage of land allocated to cotton	61 (all Bt)	63 (90 per cent Bt; 10 per cent NPM)	17 (all organic)
Percentage of population landless	5 (21 households)	10 (97 households)	0.5 (1 household)

FC Forward Caste, *BC* Backward Caste, *SC* Scheduled Caste

It is evident from Table 6.1 and details of the village composition (see Appendices 4, 5 and 6) that the ascription of dominance to particular castes coincides with that caste's relative land ownership and numerical presence within the village. It also involves considerations of ritual status (hence, the Madigas are not the dominant caste in Nandanapuram despite their strong numerical presence—see Appendix 5). As will be explored, the dominant caste has a significant influence on the way in which risk is constructed and distributed in the village.

Townsend (1994: 560) highlights the importance of the 'smoothing device' of a social network in risk-sharing. Caste is not simply a source of conflict and power struggle in the villages; instead, there is also evidence of intra-caste solidarity. In larger villages, this is facilitated by the division of the village into caste wards and the grouping of castes into their own smaller communities.

Inhabitants of all villages were found to assist each other within their caste groups. This could be seen in the priority granted to members of one's own caste in the leasing of bullocks and the provision of wage labour during seasonal peaks (the demands of upper castes permitting). The caste group was also the most common source of informal borrowing which will be discussed later. In this way, intra-caste kinship sought to mitigate risk and to overcome the fragmentation of the village collective on the basis of its caste diversity. In Orgampalle, the collective approach to risk was, in large part, derived from the significant caste homogeneity of the village (all but four households are from the BC Mudhiraj caste).

There is a strong reliance on Bt cotton in both Bantala and Nandanapuram, and the cotton crop accounts for over 60 per cent of the cultivated area in both villages. In Nandanapuram, 10 per cent of the cotton crop is cultivated using NPM methods. In Orgampalle, Bt cotton is banned as part of the 'rules' of the village (see Appendix 7). Cotton is cultivated on just 17 per cent of village land, allowing for a greater variety of crops. Walker and Ryan (1990: 258) note the significance of diversification to risk negotiation in dryland agriculture, particularly if the source of risk is not covariate (i.e. does not affect all crops equally or simultaneously).

Although both Bantala and Nandanapuram have a female *sarpanch*, it was noted that their husbands took the lead in the village, particularly in

Bantala. I was directed to Pallav, the *sarpanch*'s husband, to seek consent to conduct the research in Bantala, and it was he who signed the consent form. Pallav was also active in organising and managing the discussion at the *gram sabha* meeting which I attended.

In 2008, 15 buffaloes died in Bantala after they escaped and grazed on the Bt cotton fields. (I was passed photographs of the villagers with the dead animals). Although there were visits from politicians, industry and government scientists, vets and NGOs at the time of the deaths, villagers have never been informed of the results of these investigations. The deaths are attributed by all participants to Bt cotton, an opinion which has been confirmed, in their view, by the official recommendation that animals should be prevented from grazing on the crop.

It should be noted that the animals which died in Bantala had escaped and would not have been grazing on the cotton fields as a standard practice. The fact that pesticide use has declined since the introduction of Bt cotton suggests either that there has been a change in the strength of the pesticides which farmers are using or that the deaths are due to a broader spectrum of toxicity in the Bt plant than is asserted by proponents. Either way, reports of the deaths raise concern for the farming praxis associated with Bt technology in relation to food crops.

This study adopts the land classification used by Mishra (2007: 5) and Omvedt (1994: 329)—however, this has been converted from hectares to acres as follows:

Marginal	0.1–2.5 acres
Small	2.6–5 acres
Semi-medium	5.1–10 acres
Medium	10.1–20 acres

As will be covered in the next section, participants may not allocate all of their land to cotton cultivation. When referring to the cotton land-holding of participants, this will be specified; otherwise, land-holding details can be taken to relate to the total land-holding of the participant. The particular power structure associated with each village will now be explored.

Bantala: Concentrated Power and the Legitimation of Bt Cotton

Details of the caste composition and land-holding pattern for Bantala can be found in Appendix 4. A village map with the locations of the participants' homes is also provided as Appendix 8. As highlighted in Table 6.1, power in Bantala is concentrated in the hands of a dominant BC Kuruma caste who own 51 per cent of village land and a small number of Forward Caste Reddy and Vaishya land-holders. The Scheduled Caste (SC) population is split between Malas and Madigas, the majority being Madigas (88 per cent). Despite representing 35 per cent of the total Bantala population, the SC Madigas own just 14 per cent of the village land. This includes land allocated during land reform which is unfit for cultivation.

The suppression of the Scheduled Castes in Bantala is secured by the dominant BC Kuruma caste and the Forward Castes. This is evident in the way inequality relating to caste is legitimated in the village. Chitta (male, dominant caste, semi-medium cotton-holder) argues, 'we have inherited the caste system. We all believe in it, and belong to it. There, it is written that, as fingers on the same hand are all unequal, so castes are not equal.'

The segregation of the castes in the village is also described by Ashna (female, SC Madiga, cotton small-holder). She observes, 'they [the upper castes] do not allow lower castes to enter their houses. They will also not eat food [we] prepare.' Ashna notes the way in which the caste system is defined by upper castes, claiming 'when we are born, everyone is equal, but the upper castes say it's been written that they are superior.' She also indicates the precarious legitimation of this power structure in the village, however, and the strict boundaries within which it is tolerated. She argues:

> if you have money, if you have good [fair] skin colour, if God has given that to you, I'm not asking you to give it to me. You live in your realm, and I will live in mine. But do not provoke me. If you provoke me, I will fight against you.

The suppression of the Scheduled Castes in Bantala is evident not only from their limited access to land and other assets (see Appendix 11); it is also apparent from the pay rates for agricultural labour in the village, given that, as highlighted, most daily wage labourers are from the Scheduled Castes. Daily rates rarely went above Rs 100 for females and Rs 150 for males.[2] The lower wage rates in Bantala heightened the risk exposure of the landless daily wage labour in the village, particularly females, while mitigating the risk of the dominant caste and Forward Caste landowners.

The intersection of opportunities for patronage with power structures in the village was highlighted by Nipa (female, SC Madiga, landless participant) who shares her sister's house with her husband. When asked why she hasn't been able to access land or a house through land reform and the Indiramma initiative, she replied: 'I didn't get anything because the *sarpanch* didn't write anything for me.' As highlighted in Chap. 3, access to resources relies upon a written request of potential beneficiaries to be passed through the PRI tiers. Illiterate villagers are reliant upon others, particularly the *sarpanch*, to initiate this request on their behalf. Nonetheless, Nipa legitimates the power structure given her perception that things are generally improving as a result of it. She claims, 'last year we used to get 50–60 rupees per day [for daily wage labour]. Now we get 100.'

Nandanapuram: Contested Power and Legitimation Crisis

The village composition for Nandanapuram is included as Appendix 5. The village map and residence locations of Nandanapuram participants are provided as Appendix 9. An NPM centre, which dispenses the solutions required to undertake NPM cultivation and offers advice on the method, is located in the centre of the village.

As Appendix 5 highlights, land-holding in Nandanapuram is relatively evenly spread across a number of influential castes, including a highly mobilised Scheduled Caste population. The Scheduled Castes in

Nandanapuram are comprised entirely of Madigas who own the majority of the village land (37 per cent) and represent 41 per cent of the population. Their low caste status and employment as daily wage labour means, however, that they are not identified as dominant. Instead, a number of powerful groups, including the FC Reddys and BC Gowdas, as well as the SC Madigas, compete for power in the village.

Ranjan (male, BC Chakali, landless) describes the fraught power relations in the village. He claims, 'the Madigas don't even fear the Reddys or Brahmins. The Reddys need the Madigas for work. They will pay double or treble the going rate and never raise their voice against them.'

The delegitimation of traditional caste relations in Nandanapuram, and the strength of the Scheduled Caste population, is confirmed by Nikhil, an FC Reddy participant. He claims:

> during my grandfather's time, there used to be respect for the Reddys as a higher caste. Everyone used to listen to us. But now not even a higher-caste status is given to us. No-one from the lower castes even cares what we say. They say now we are all equal. (Nikhil: male, FC Reddy, medium land-holder)

Charan (male, FC Brahmin, semi-medium cotton cultivator), an elderly Brahmin, also states, 'We don't get respect [as Brahmins]. That was our time. This is not our time.'

The sense of anomie in Nandanapuram arising from the contested power structure is captured well by Ranjan (male, BC Chakali, landless). He observes, 'there are a lot of limitations about who to talk with, and who not to talk to. It's like standing at a crossroads when you don't know where to go, or how to behave.' He notes how the changing power dynamics have altered everyday village practices, observing 'earlier no-one used to take water from Madigas. Now the Madiga works on the water tank and distributes water.'

The relative power of the SC Madigas in Nandanapuram, the majority of whom are wage labourers, is evident from the fact that daily wage rates in Nandanapuram were higher than in the other two villages. Participants reported that up to Rs 150 was paid for females in the village and Rs 250 for males at peak times, such as harvesting. While the higher rates alleviated the risk exposure of agricultural labourers, especially males, they heightened that of cultivators in the village, particularly the marginal and small-holders whose ability to pay the high labour costs was more limited.

Orgampalle: Charismatic Power and Delegitimation of Bt Cotton

The composition of the village of Orgampalle is provided as Appendix 6. The locations of participants' homes on the village map are included as Appendix 10. For administrative purposes, Orgampalle is linked to a neighbouring village where the *sarpanch* (a male, BC Mannuru Kappu) is based. Within Orgampalle, however, Pradnesh (male, dominant caste, medium land-holder), a village elder, is the most powerful villager.

Pradnesh is much revered and highly influential in promoting a co-operative approach in the village. This is facilitated by the fact that all but four households (two BC Yadava and two BC Chakali) are from the dominant BC Mudhiraj caste, whose members also own 82 per cent of village land. An organic centre is located at one end of the village. On a board on the wall, records of ground-water levels and details of the occurrences of pests and diseases, as well as the organic solutions used to treat these, are maintained throughout the season.

Aruni (female, dominant caste, marginal cotton cultivator) highlights the absence of caste conflict in Orgampalle. She claims, 'we go to each other's houses. We don't have any caste feelings here.' And Achanda (male, BC Chakali, cotton small-holder) notes, '[a]ll castes share the same temple' (although the village has two temples, their use is associated with different festivals).

Pradnesh is aware that his leadership of Orgampalle operates against the prevailing neoliberal philosophy of the pre-secession state. He claims, 'compared to outside [society beyond the village], we may be backward in terms of our development. But my philosophy is, let them do what they want. They may be going in circles, but let's stick to our own path.' Pradnesh emphasises a more collective approach as the means to mitigating risk exposure generally. He observes, 'we help each other even in terms of sharing water, and animals [oxen for ploughing].'

The co-operative approach to risk negotiation in Orgampalle was evident from the fact that daily wage labour rates rarely rose above Rs 70. Here, the aim was to minimise the risk exposure for all cultivators equally. This was facilitated by the fact that there was no household in Orgampalle which relied upon daily wage labour. The inhabitants of the only landless household in the village were an elderly couple in poor health who received a meagre state pension for their subsistence (Rs 200 per month).

The strategy with regard to labour in Orgampalle was explained by Nirmal (male, BC Yadava, cotton small-holder):

> In this village, all people have their own lands. They will finish work on their own land first and then move on to help another. So there is mutual understanding. That's why they are paying low wage rates.

Participants are obliged to negotiate risk from within the 'social risk position' (Beck 1992: 40) associated with their land-holding, caste and gender within each village. The risk positions of participants will now be explored.

Participant Risk Profile

A total of 26 participants took part in the research, 5 of whom were landless. The spread of participants across the villages is as follows (Table 6.2):

Of the five landless participants, three are female (one widow), three are Backward Caste and two are Scheduled Caste.

The spread of the social risk positions of cultivating participants is as follows:

Table 6.2 Village participants

Village	Cultivators[a]	Landless
Bantala	6 (all Bt cotton cultivators)	2
Nandanapuram	8 (6 Bt cotton and 2 NPM cultivators)	2
Orgampalle	7 (all organic cotton cultivators)	1
TOTAL	**21**	**5**

[a]This study adopts the term 'cultivator' to refer to those who operate their own (leased or personally-owned) holdings and 'daily wage labour' for those who work on the land of others.

Table 6.3 Land-holding, caste and gender of cultivating participants

		% of participants
Land-holding	Marginal (0.1–2.5 acres)	43
	Small (2.6–5 acres)	33
	Semi-medium (5.1–10 acres)	14
	Medium (10.1–20 acres)	10
Caste	Forward Caste	20
	Backward Caste	56
	Scheduled Caste	24
Gender	Male	68
	Female	32

As Table 6.3 highlights, many of the participants can be regarded as among the most vulnerable in terms of agrarian risk according to the profile of cultivators who committed suicide in Telangana between 2003 and 2004 analysed by Revathi (2009: 217). Seventy-six per cent operate marginal and small-holdings, 56 per cent are Backward Caste and 68 per cent are male. As noted, however, the profile of farmers at risk of suicide is changing to increasingly include land-holders of up to 25 acres, as well as females (Goyal 2015: 283; x).

Appendices 11, 12 and 13 provide details of access to key assets (tractors, borewells and oxen) and off-farm employment, as well as land-holding of cotton as a proportion of total land-holding for individual participants in each village as part of their risk profile. In a semi-arid region, such as Telangana, access to irrigation is key to the mitigation of risk. As Walker and Ryan (1990: 41) note, in dryland areas, 'the incentive to own wells is stronger than the ambition to own land.'

Tractor and oxen ownership reduces the costs of cultivation. This is due to the fact that participants without such assets are required to hire oxen at an additional cost in order to plough the land to prepare it for sowing and to deal with more widespread weeds during the season. The significance of such ownership is also noted by Townsend (1994: 585) who claims that 'wealth is related to landholdings and owned bullocks.'

Appendix 11 highlights that, in Bantala, ownership of assets is concentrated among the most powerful Forward Caste and dominant caste participants. Both Pavan (male, FC Vaishya, medium cotton-holder) and Chitta (male, dominant caste, semi-medium cotton-holder) own oxen, while Sudhakar (male, FC Reddy, cotton small-holder) owns a tractor. Access to irrigation is restricted to these three participants. Natesh (male, BC Gowda, cotton small-holder) did own oxen but was obliged to sell them to pay for labour costs during the 2010/2011 season.

In Nandanapuram, as Appendix 12 indicates, while land ownership is concentrated among Forward Caste participants, particularly Nikhil (male, FC Reddy, semi-medium cotton-holder) and Charan (male, FC Brahmin, semi-medium cotton-holder), ownership of other assets is slightly more diffuse than in Bantala. Rajiv (male, BC Gowda, cotton small-holder) owns a borewell, and Ambu (female, SC Madiga, cotton small-holder) owns oxen and shares a borewell with her deceased husband's brothers. Ambu is the only Scheduled Caste female in the study to own oxen or to have access to irrigation.

Appendix 13 provides details of asset ownership in Orgampalle. Borewells are banned in Orgampalle given their impact on ground-water levels. One participant (Achanda: male, BC Chakali, cotton small-holder) has a borewell which he drilled prior to the ban. Its use is restricted to his cultivation of paddy which requires more water than cotton. Land ownership is concentrated in the case of Orgampalle participants with the village elder, Pradnesh, being the sole medium land-holder; the ownership of oxen, however, is far more diffuse than in the other villages. Amita (female, BC Chakali, marginal cotton cultivator), a widow, is the only participant in Orgampalle who does not possess oxen. This means that for participants in Orgampalle, the main cultivation cost is generally that associated with labour.

Appendices 11, 12 and 13 provide details of any off-farm income to which participants have access. It also indicates those with additional land-holding which is not allocated to cotton. As highlighted, the diversification of crops can alleviate exposure to risk and allow some compensation for the loss of a particular crop in the event of non-covariate risk. Given that the focus here is on Bt cotton cultivation, the analysis does not provide details of the impact of the cultivation of other crops on the household's overall exposure to risk. However, the proportion of the total holding allocated to cotton provides an indication of the significance of cotton to the risk negotiation of particular households.

As discussed, off-farm employment is an important aspect of risk negotiation in rural areas given that income from such employment can subsidise poor crop years. Access to employment is also important when the agricultural season has ended. Walker and Ryan (1990: 260) also note the importance of off-farm employment to alleviating agrarian risk, particularly with regard to self-targetting rural public works programmes.

The chances of obtaining well-paid off-farm employment are greatly increased with education. Education, however, also involves risk in the rural context given the impact of the loss of unpaid agricultural workers on the labour costs which must be borne by the household. Failure to obtain employment following education means that the household sacrifice of unpaid agricultural labourers to enable attendance at school and college, and the additional cultivation expenditure which this has given rise to, has been futile. There is also the loss of income associated with school and college attendance given that many cultivators work as wage labour on the land of other villagers in order to fund their own cultivation costs.

Details of off-farm employment among participants indicate a variety of activities. These include construction (in Bantala, both female, SC Madiga participants have husbands who work in construction in a nearby town); NGO employment (both Nand, a BC Yadava in Nandanapuram and Prakash, a BC Yadava in Orgampalle, work for local NGOs); auto-rickshaw owner and driver (Sudeep, an SC Mala in Bantala); and director of a co-operative bank (Chitta, dominant caste, semi-medium cotton-holder in Bantala).

There is also evidence of traditional caste occupations such as that of priest (Charan, FC Brahmin in Nandanapuram); shop and rice mill owner (Pavan, FC Vaishya in Bantala); laundry work (Ranjan, BC Chakali in Nandanapuram) and toddy tapping (the son of Natesh, BC Gowda, in Bantala). Given the absence of Brahmins in Bantala, an FC Reddy (Sudhakar) has taken on the role of village priest. This is frowned upon in the village, with one dominant caste female noting disparagingly that Sudhakar's only claim to priest-hood is that his ancestors paid to build the temple.

All of the participants with access to off-farm incomes, including those associated with caste occupations, are male. This suggests that the opportunities for rural females to access off-farm employment are far more limited than for males.

Details of land lease per participant can also be found in Appendices 11, 12 and 13. As highlighted, the leasing of land represents an additional risk given that the lease rent is payable if the crop fails. Six of the 21 cultivators leased land. Four of these are located in Bantala; of these, two are Forward Castes: Pavan (male, FC Vaishya) and Sudhakar (male, FC Reddy) and one, Chitta, is from the dominant BC Kuruma caste. The fourth is Natesh, a BC Gowda village elder.

The study suggests that land lease is often arranged between castes. In Bantala, both Pavan (FC Vaishya) and Natesh (BC Gowda) lease land from an FC Reddy who lives in the United States,[3] while Sudhakar's (FC Reddy) land lease was arranged with a BC Gowda who operates a business (unspecified, despite questioning) in the village. The prevalence of leasing among the more powerful Bantala participants suggests the legitimation of risk-taking as the means to development in the village.

In Nandanapuram and Orgampalle, only one participant leases land in each village. In Orgampalle, Achanda (BC Chakali) leased land from a dominant caste BC Mudhiraj; in Nandanapuram, Rajiv (BC Gowda) leased village land from a relative (same caste) living in another Telangana district. Both of these villagers can be regarded as relatively powerful in their respective villages. They both own borewells, and Achanda receives remittances from a son working in a nearby town. Rajiv shows a tendency towards risk-taking given that he introduced Bt cotton to Nandanapuram. He is also, as will be explored, significantly in debt.

There have been some changes among participants in terms of their access to land. A number of the older participants, particularly in Nandanapuram, noted that they had bought land from *doras* (local landlords) following the armed struggle. Sudeep (male, SC Mala, marginal cotton-holder) in Bantala also recounted how his father had been in bonded labour with a local landlord for 20 years before buying land with savings he had managed to put aside over time.

There is also evidence of distress sales of land. Nipa (female, SC Madiga, landless) noted that her father was obliged to sell his land in order to pay for the dowry of his four daughters. And Natesh (male, BC Gowda, cotton small-holder) told how he had considered suicide a few years earlier because of the indebtedness associated with dowry costs for his sisters and daughter, as well as crop loss. Instead, he sold 6 acres of his land.

The social risk position associated with caste was found to be particularly important in the ability to access village labour. Demand for daily wage labour is highly seasonal with peak times associated with sowing, weeding (the degree and timing of which depends upon the rains) and harvesting. Because everyone in the village required wage labour at the same time, villagers were often obliged to make use of labour from other villages. This labour was more expensive as cultivators were required to pay the transport charges for these workers which could be anything up to an additional Rs 50, depending upon the journey.

While most participants noted their need to make use of the more expensive labour from other villages, both upper caste FC Reddy males in Bantala and Nandanapuram (Sudhakar and Nikhil, respectively) and the FC Brahmin male (Charan) in Nandanapuram claimed that they used labour only from their own villages. Thus, unlike the other participants, they did not have to incur the extra costs associated with hiring workers from other villages during the peak season. In Orgampalle, participants did not make use of labour from outside the village but co-ordinated their availability to ensure that peaks in demand were covered amongst themselves.

This chapter will now explore how these power relations and risk profiles influence the way Bt cotton is legitimated and delegitimated in the villages. This is undertaken through the use of the prestige and conformist biases developed by Henrich (2001: 997) and first applied by Stone (2007: 71) to Bt cotton in Warangal.

Bt Cotton and Risk Negotiation in the Villages

Figure 6.1 provides an overview of the differing perspectives among village participants regarding Bt cotton. It can be seen that the majority of participants strongly legitimate Bt cotton in Bantala. In Orgampalle and Nandanapuram, however, the technology is delegitimated, with the majority of participants strongly disagreeing with the statement 'I want to grow Bt cotton.'

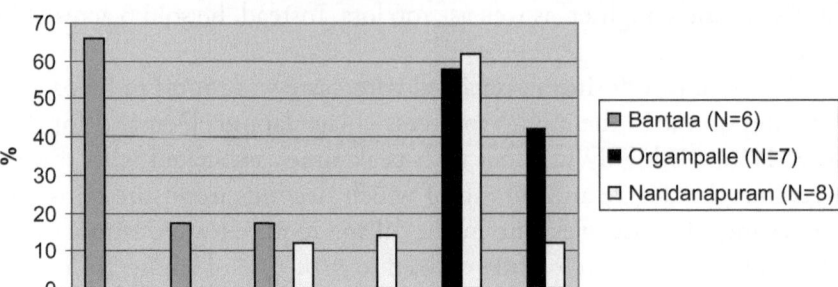

Fig. 6.1 An overview of village perspectives on Bt cotton

A small number of Bt cotton cultivators in Bantala and Nandanapuram are uncertain (the 'don't know' category) as to whether they wish to cultivate Bt cotton given their assertions of declining yields. There are also a number of participants in Nandanapuram and Orgampalle who are unaware as to what Bt cotton is. (This was despite trying different naming conventions, such as Bollgard, to check understanding).

In Nandanapuram, one female participant (Rashi, FC Brahmin, marginal cotton-holder) said she did not know what Bt cotton was and just followed her neighbours. The lack of awareness of Bt cotton is particularly prominent in Orgampalle, where three of the seven cultivators claimed not to know anything about Bt cotton. They comprised the two female participants and Nirmal (male, BC Yadava, cotton small-holder). The limited knowledge concerning Bt cotton in Orgampalle suggested the reliance, as expressed by a number of participants, on Pradnesh and Crops Jangaon to make decisions on their behalf.

The situation with regard to the perceived legitimacy of Bt brinjal was, however, very different, particularly in Bantala. As Fig. 6.2 highlights, Bt brinjal was delegitimated in all three villages, with the majority of participants strongly disagreeing with the statement that they would have no health concerns about eating Bt brinjal.

Q. 7 I would have no health concerns about eating Bt brinjal

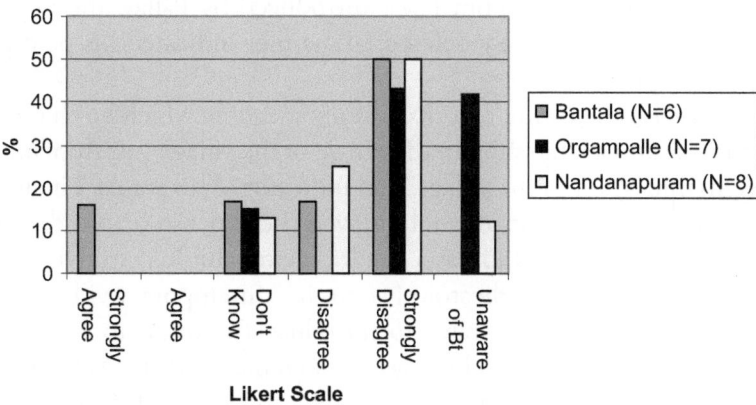

Fig. 6.2   An overview of village perspectives on Bt brinjal

Only one of the participants in Bantala strongly agreed that they would have no health concerns in this regard (Chitta: male, dominant caste, semi-medium land-holder). Generally, however, the marked change in perspective when compared with Bt cotton was asserted to be linked to the view that Bt brinjal was a food crop (despite the fact that cottonseed oil is widely used in cooking). The belief of Bantala cultivators that Bt cotton was responsible for animal deaths contributed to their concern regarding the use of Bt technology in a food crop.

The impact of village power relations on the legitimation (and delegitimation) of Bt cotton will now be explored through the use of the prestige and conformist biases identified by Henrich (2001: 997). This seeks to highlight the way in which the legitimation of risk, as a concern with the reasons for the adoption (or rejection) of an ambiguous technology in an already high-risk context, is mediated through power relations.

Prestige Bias

Pavan, the FC Vaishya, is the most powerful participant in Bantala. He grows cotton on 20 acres of land (17 of which is leased). His elevated status in the village was evident not only from his dismissive attitude to the Scheduled Castes but also from a number of other aspects. Pavan was the first participant to whom I was introduced by Pallav, the *sarpanch's* husband. Pallav's unusually deferential manner indicated his perception of Pavan's superiority.

Pavan is also the owner of a rice milling machine which he charges villagers to use and a large shop in the centre of the village (see Appendix 8). His central location in the village is an indicator of his status. Finally, one of his sons is studying in Australia (given the expense involved and the likelihood of a well-paid job as a result, having children studying in the US or UK is a primary indicator of status in contemporary village India).

Pavan was the first cultivator to introduce Bt cotton to Bantala. He claims he heard about the technology through the 'market.' 'Market,' 'market rate' and 'money' are some of the limited English words which all cultivating participants in Bantala, regardless of caste or land-holding, know. Pavan was told about Bt cotton by seed dealers and began growing the crop on a few acres. When other villagers noticed the higher yields which he achieved, they followed. Pavan grows Bt cotton because he claims 'there has been a 25 per cent increase in income with Bt cotton' and argues, 'because fewer pesticide sprays are required, Bt cotton is better for the land.'

The contested power relations in Nandanapuram, and the competing authorities which this gives rise to, are evidenced in the struggle to secure stable legitimation for any one cultivation method in the village. Bt cotton was first introduced to the village by Rajiv, a male small-holder from the BC Gowda caste. As highlighted, BC Gowdas are relatively powerful in Nandanapuram, representing 20 per cent of the village population and owning 26 per cent of the village land (see Appendix 5). They are ranked third in the village in terms of numerical presence and land-holding, after the SC Madigas and the Boyas (a Scheduled Tribe). Rajiv explains how his introduction of Bt cotton (prior to the 'official' approval of the crop) occurred:

> Twelve years ago, I went to [a nearby town]. I knew a trader there, so I asked him for two packets of Bt cotton seeds [sufficient for two acres]. I gave one packet away because the ladies of the household were afraid, saying 'you can't eat food in the field if it's a Bt crop' [cultivators eat their lunch in the fields]. But I kept one bag and tried it. I got a good yield and everyone copied. (Rajiv, male, BC Gowda, small-holder)

While Pavan continues to legitimate Bt cotton in Bantala, however, Rajiv's ongoing adoption of Bt cotton now coincides with his active campaigning against it. Rajiv (male, BC Gowda, cotton small-holder) argues, 'Bt crop yields have declined, but my expenditure keeps increasing because of pesticide use.' His ongoing need for pesticides is a cause of concern for him not only given the expense but also because of his fears for its sustainability. He states, 'soil fertility will be lost so future generations will have nothing to eat.' Despite being the first to introduce the technology to the village, therefore, Rajiv now claims 'to be following others.'

The delegitimation of Bt cotton in Nandanapuram by cultivators who continue to adopt it is in many ways embodied in the conflicted position of Rajiv. He continues to cultivate the technology, even as he speaks at mass rallies of farmers in Warangal against the use of Bt cotton and pesticides, and demands that the government make alternative options available. As will be highlighted, Rajiv's ability to adopt the lower cost, but also potentially lower yielding, NPM alternative which is available to him in the village, is restricted given his significant debt levels. (In fact, in the current study, it was Nand, an NPM cultivator, who achieved the highest yield of nine quintals per acre).[4]

Despite the differing views on Bt cotton in Bantala and Nandanapuram, the introduction of Bt cotton followed a similar path where influential villagers were encouraged by seed dealers to try the new varieties. (It is not clear whether these villagers were specifically targeted by seed dealers who were aware of their influence). Given the status of these villagers, and the improved yields witnessed by the other villagers, the technology quickly diffused within the villages.

The competing presence of an alternative cultivation method in Nandanapuram, however, meant that Bt cotton was not adopted uniformly by all cultivators. Instead, a small number of villagers in Nandanapuram (10 per cent) opt for Non-Pesticide Management (NPM), a cultivation practice in which Bt seed varieties are prohibited. In Nandanapuram, this method is promoted by Nand (male, BC Yadava, cotton small-holder).

Nand is in his 30s and soft-spoken. He is university-educated (he has a degree in Sociology), has won awards for his agricultural practice and is an NGO employee (SEED). These are all sources of prestige in the village.

His elevated position was evident from the fact that my translator (an SC Madiga) used Telugu's polite form (*meeru*) to address him. Nand actively campaigns against Bt technology and promotes NPM practices in which the use of pesticides and Bt cotton seed varieties are banned (because of the belief that the pesticide is in the Bt cotton plant).

Despite these obvious signs of prestige, however, Nand's caste, the BC Yadavas, is not particularly powerful in Nandanapuram (see Appendix 5). They are numerically poorly represented (they account for just 2 per cent of the village population) and have limited access to village land (2.6 per cent). The relatively limited power associated with Nand's caste status in the village, as well as the debt levels among cultivators in Nandanapuram which will be explored in the next section, contribute to the fact that, despite Nand's individual standing, only a small number of cultivators have opted to follow his lead in their cultivation practice.

Nand strongly delegitimates the adoption of Bt cotton. He argues, 'the height of the crop has been decreasing…[leading to declining yields, and] the quality of the soil is deteriorating day by day [due to pesticide use].' He also claims that the bollworm has developed resistance and argues that 'sucking pests have doubled.' This means that pesticide use is again increasing. Nand claims, 'the government is not giving an equal chance for Bt and non-Bt cultivation methods' and argues, 'the MNCs [multinational corporations] who produce Bt cotton influence the government.'

As highlighted, in Orgampalle, Pradnesh is an influential and much respected village elder. He is from the dominant BC Mudhiraj caste and a landowner of 20 acres. He strongly advocates the organic methods now adopted by all villagers. According to Prakash (male, BC Yadava, marginal land-holder), up until 2003, a number of cultivators in Orgampalle used pesticides; however, when they noticed that they were obtaining the same yields as organic cultivators but were spending more, they followed the lead of Pradnesh and switched to organic cultivation.

The authority of Pradnesh in the village is secured through the fact that his ancestors were among the first to settle there. He claims to have become aware of the risks of Bt cotton through his interaction with cultivators in other villages. Pradnesh then used his influence to persuade all villagers in Orgampalle, in conjunction with Crops Jangaon, that Bt cotton should be avoided. He now actively campaigns against the technology. He says, 'I am telling cultivators in meetings not to grow Bt varieties.'

Pradnesh argues that 'cultivators are being cheated by the seed companies.' He claims, 'Bt cotton seems to be increasing wealth for some time, but it will deteriorate sooner or later.' He also believes that the Bt plant is toxic, arguing 'chemicals have poisoned the seed and that is carried in the plant.' This, he claims, is leading to 'animals dying which graze on the fields. There is also a problem of skin allergies if we work on the Bt cotton fields.'[5]

Conformist Bias

Conformist bias is particularly evident among the more vulnerable participants. A number of the female Bt cotton cultivators in Bantala and Nandanapuram claim to adopt the seed variety which the majority are opting for. Rashi (female, FC Brahmin, marginal cotton cultivator) in Nandanapuram claims, 'If 10 of my neighbours go to one shopkeeper, I'll see which variety the majority opt for, and then choose the same.' Likewise, Ashna (female, SC Madiga, marginal cotton cultivator) in Bantala states, 'everybody is growing Bt cotton, so I am also growing it.'

Both are aware of the financial risk of Bt cotton adoption. Rashi claims, 'When it's luck, I get profit. When it's unfortunate, I don't get anything. It's only luck.' And Ashna states, 'it's pure luck. If you get the crop, you benefit. Otherwise you lose.' The risk is, however, legitimated by virtue of the fact that others have opted for the same risk and have at times obtained high yields.

All land-holding classes of male cultivator also indicate evidence of conformist bias in their adoption of Bt cotton. Sudeep (male, SC Mala, marginal cotton-holder) in Bantala states, 'because others were getting more yields with Bt, I also tried it.' Likewise, Charan (male, FC Brahmin, semi-medium cotton-holder) in Nandanapuram claims, 'I have seen the other cultivators who have grown it. It gives them higher yields.'

The risk of Bt cotton is legitimated in Bantala despite the awareness of the concerns in wider Warangal society regarding the technology. Sudhakar (male, FC Reddy, semi-medium land-holder) argues, 'the non-Bt fellow will say Bt is harmful. We listen and that's it. We don't know if it's really harmful or not.' This legitimation arising from the uncertainty of the risks involved is asserted despite this participant's belief that the technology may be responsible for the deaths of animals as in his claim: 'once buffaloes ate Bt cotton and 15 of them died.'

Despite evidence of conformist bias, the ambiguity of participants in Bantala regarding the sustainability of their cultivation practice is asserted by Chitta (male, dominant caste, semi-medium land-holder) who claims, 'I heard the land is getting spoiled, so in the future we may not get good yields.' This leads him to question, '[e]verybody grows cotton. But tomorrow what will happen?'

Bt cotton is legitimated in Bantala, despite these concerns, given the assertions by all participants of the higher yields and increased profits which Bt cotton has made possible, as well as their limited alternatives. Natesh (male, BC Gowda, cotton small-holder) argues, 'we're growing Bt cotton because if we go for groundnut or something else, insects…will attack and we'll incur loss.' He also notes that pesticide use has declined. He claims, 'I used to spend more on pesticides.'

Sudeep (male, SC Mala, marginal cotton cultivator) claims that vegetable crops are prone to other predators. He argues, 'there are a lot of monkeys, so there is no safety in growing vegetables.' And Pavan (male, FC Vaishya, medium cotton cultivator) notes, 'earlier we used to get 5 or 6 quintals. Now we're getting 12.' This, he asserts, has led to a 25 per cent increase in income.

Both low caste cultivators in Bantala assert the increased profitability of Bt cotton. Ashna (female, SC Madiga, small-holder) claims, 'we are getting more yield, so we are getting more money.' Likewise, Sudeep (male, SC Mala, marginal cotton cultivator) asserts, 'financially we benefitted. We're sending my brother to school, we've built a house and we're not relying on others for loans.'

In Nandanapuram, Nikhil (male, FC Reddy, semi-medium cottonholder) also asserts the limitations in crop choice caused by water shortage, claiming 'I tried crop rotation on one part of my land, but I couldn't try it elsewhere because there is no water.' He observes, 'if everyone is growing Bt cotton and making money, then I must also do it to live. I'm just following the blind.' He is also aware, however, that not everyone can withstand the risk involved. He claims, 'a lot of cultivators can't bear the losses so…they have to sell land in order to clear debts.'

As highlighted, the contested power structure in Nandanapuram has contributed to a mixture of cultivation methods being legitimated in the village. The delegitimation of Bt cotton even by many of its adopters is influenced by the conflicted position of Rajiv (male, BC Gowda, cotton small-holder) who introduced Bt cotton to the village and who continues to cultivate it even while he actively campaigns against it.

Conformist bias is also apparent in the rejection of Bt cotton, however. Although he is held in high esteem in the village, the NPM cultivator, Nand (male, BC Yadava, semi-medium land-holder), has been unable to influence more than a small minority to adopt NPM practices. This is despite evidence of the delegitimation of Bt cotton in the village (see Fig. 6.1). While the status of the Yadava caste in Nandanapuram is relatively low (see Appendix 5), the fear of changing cultivation practice in Nandanapuram is also influenced, as will be explored, by the debt levels of participants. Nonetheless, Nishok (male, SC Madiga, marginal cotton cultivator) asserts Nand's influence on his own adoption of NPM practices. He claims, 'it is only through [Nand], that I came to know about [NPM].'

In Orgampalle, the influence of Pradnesh in the delegitimation of Bt cotton is noted by all participants. Achanda (male, BC Chakali, cotton small-holder) asserts, 'until now, we are following Pradnesh's wishes.… The people accept his seniority.' Many cultivators have, however, also rationalised the ban on Bt cotton in the village in their own terms. Nirmal (male, BC Yadava, cotton small-holder) claims, 'cultivators commit suicide because they are paying for chemicals.…The production cost is Rs 20,000 per acre. It's very difficult to survive.'

The role of Crops Jangaon, whose influence is legitimated by Pradnesh, is also noted as crucial. Achanda (male, BC Chakali, cotton small-holder) asserts, 'we would check with Crops [Jangaon] for their advice [with regard to cultivation decisions].' Akhil (male, dominant caste, marginal cotton cultivator) claims, 'the debt problem has eased since [Crops Jangaon] started working with us. They want to create this village as a role model for the entire country.' For others, however, the rules are not so clear. Achanda (male, BC Chakali, small-holder) claims, '[Crops Jangaon] won't allow anyone to buy Bt cotton. They supply the [non-Bt] seeds. I don't know why they object to Bt cotton.'

The female participants in Orgampalle follow the advice of others in their approach to cultivation, particularly elders. Amita (female, BC Chakali, marginal cotton cultivator) states, 'I'm not aware of anything to do with Bt cotton.' And Aruni (female, dominant caste, marginal cotton cultivator) claims, 'the village elders know.'

The debt exposure of participants will now be explored. This seeks to examine the impact of the relative power of participants on their differentiated exposure to risk and to explore the effect of Bt cotton on this.

Debt Exposure and the Legitimation of Risk

Given that indebtedness is noted as an important factor with regard to farmer suicides (Vasavi 2012: 22–23), the analysis here seeks to examine the way in which indebtedness intersects with power relations in the legitimation of risk. Here, the study seeks to explore the contribution of the additional costs, but also potentially higher yields of Bt cotton, to the risk exposure associated with indebtedness over time.

Full details of the debt levels of each participant, as well as the sources and reasons of borrowing, are included in Appendices 14, 15 and 16. These are also summarised in Table 6.4. Figures here represent average accumulated debt levels of participants including debt associated with the 2010/2011 season.

As Table 6.4 highlights, the average debt levels of Bt cotton cultivators in both Bantala (Rs 227,500) and Nandanapuram (Rs 191,666) are significantly higher than those for NPM cultivators in Nandanapuram (Rs 12,500) and organic cultivators in Orgampalle (Rs 48,428). Interestingly, average debt levels are highest for cultivators in Bantala, a village in which Bt cotton is strongly legitimated given its contribution to higher incomes and yields, and reduced pesticide expenditure (see Fig. 6.1). It was this positive assessment of the contribution of the technology to incomes which had been traded against the concerns of participants for the future sustainability of Bt cotton and its possible links to animal deaths.

Table 6.4 Average accumulated debt levels per village, reasons for and sources of debt by land-holding of cotton, caste and gender

Cultivator numbers	BT	NP (Bt)	NP (NPM)	OR	Reasons for debt	Sources of debt
	N = 6	N = 6	N = 2	N = 8		
Average debt levels (excl. landless) (Rupees)[a]	227,500	191,666	12,500	48,428		
Average debt level per land-holding of cotton (excluding the landless)						
Marginal (0.1–2.5 acres)	25,000	215,000	25,000	47,250	Cultivation costs; dowry; health; input and labour charges; house construction; borewell	Bank; MFI; trader in other villages; people from own or nearby villages and/or own caste; NGO society
Small (2.6–5 acres)	382,500	180,000	0	50,000	Buying land; tractor; borewell; shed; dowry; cultivation costs; health; labour; education; house construction	Bank; people in own and other villages; people from other castes; family members
Semi-medium (5.1–10 acres)	150,000	110,000	—	—	Input and labour costs; cultivation costs; education	Banks; private money lender in another village; people in own villages
Medium (10.1–20 acres)	400,000	250,000	—	—	Cultivation costs (seeds, inputs, labour)	Banks; people in other villages; commission agent
Landless	62,000	10,000	—	—	Health; poverty reduction project; lease of land	Others in village; family; landlords; bank (landowner as guarantor); Self-Help Group

(continued)

Table 6.4 (continued)

	BT	NP (Bt)	NP (NPM)	OR	Reasons for debt	Sources of debt
Average debt level per caste (including the landless)						
FC	**550,000**	**236,667**	—	—	Cultivation costs; land; borewell; tractor; shed; education; house construction; dowry	Banks; people in own or other villages; commission agent
BC	105,000	110,000	0	42,375	Borewell; cultivation costs; dowry; health; house construction; labour charges	Banks; private lender in other villages; people in own and other villages; trader in another village; family; NGO society; MFI Bank; MFI; landlords; bank (landowner as guarantor); people from own and other villages; family
SC	24,667	60,000	25,000	—		
Average debt level per gender (including the landless)						
Male	**263,000**	138,000	12,500	31,500	Cultivation costs; buying land; tractor; shed; borewell; dowry; health; education; house construction	Banks; people in own and other villages; private money lender in other villages; commission agent; trader; family; NGO society
Female	58,000	**156,667**	—	**75,000**	Cultivation costs; health; poverty reduction; house construction; dowry; education; land lease; borewell	Bank; MFI; Bank (landowner as guarantor); landlords; people in own and other villages; family; Self-Help Group

BT Bantala, *NP (Bt)* Nandanapuram Bt cultivators, *NP (NPM)* Nandanapuram NPM cultivators, *OR* Orgampalle (organic)

[a]Given attempts to analyse the likely contribution of Bt cotton cultivation to indebtedness, the debt exposure of the landless participants has been treated separately for land-holding (unlike in the thesis from which the data is taken). The landless have, however, been included in calculating the average debt levels by gender and caste to highlight the risk associated with these dimensions arising from their limited access to resources

The breakdown in Table 6.4 indicates that male, Forward Caste medium cotton cultivators in Bantala have the highest debt exposure of the study (Rs 550,000 for Forward Castes and Rs 400,000 for medium land-holders). They are closely followed by small-holders in the village (Rs 382,500). Analysis at the individual level shows that Pavan (male, FC Vaishya, medium cotton cultivator), who introduced the technology to the village, has the second highest individual debt exposure of the study (he is Rs 400,000 in debt). The highest is Sudhakar (male, FC Reddy, semi-medium land-holder), also in Bantala, who has an individual debt exposure of Rs 700,000 (see Appendix 14).

As Appendix 14 also highlights, the reasons cited for the indebtedness of these Forward Caste Bantala participants include leasing and cultivation costs (seeds, inputs and labour), dowry and the education of children (Pavan), the purchase of land and a tractor, and the drilling for a borewell (Sudhakar). It could be argued that this indebtedness is progressive given that it involves attempts to improve the participant's life chances and/or to alleviate future risk (e.g. the education of children, the buying of land and a tractor). However, this outward display of success among Forward Castes serves not only to legitimate their upper caste status; it also legitimates their cultivation practice, particularly in relation to more vulnerable lower castes. This is despite the fact that these displays of wealth are being funded through borrowing.

The highly indebted Forward Caste participants in Bantala have assets such as land to sell should they be required to do so. Pavan owns a relatively small amount of land (3 acres; he leases a further 17) but operates a successful village shop and a rice milling machine; Sudhakar has a tractor and owns a total of 12 acres of land (an additional 4 acres are leased). Both also enjoy the social status associated with their Forward Caste ranking.

Those who seek to emulate their lifestyles, however, most notably the Scheduled Caste small and marginal cultivators, possess small tracts of land which are sufficient to secure loans; however, their vulnerability to pauperisation and landlessness is far greater given their more limited asset base and their low status within Indian society. For the landless, emulation of Forward Castes is out of the question given their limited ability to access loans as a result of their lack of possessions to offer as collateral for borrowing from institutional sources.

As can be seen from Appendices 14, 15 and 16, only 5 of the 26 participants in the study were debt-free. These included Sudeep (male, SC Mala, marginal cotton cultivator and auto-rickshaw owner) in Bantala; Nand (male, BC Yadava, cotton small-holder and NGO employee) and Ranjan (male, BC Chakali, landless) in Nandanapuram; and Pradnesh (male, dominant caste, cotton small-holder) and Sajan (male, dominant caste, landless) in Orgampalle. Both landless participants noted that their debt-free status was not necessarily through choice. As Sajan, an elderly pensioner in ill health, asked: 'Who would lend to me?'

Table 6.4 highlights that, apart from the marginal land-holders in Bantala, Bt cotton cultivators with marginal and small-holdings have a significantly higher average debt exposure than those in the same categories adopting organic and NPM methods. The debt exposure for Bt cotton cultivators with small-holdings in Bantala is almost as high as for the medium land-holder (Pavan), whose cultivation practice they are emulating.

The marginal cotton cultivators in Bantala, Sudeep (male, SC Mala) and Ashna (female, SC Madiga), have a low average debt exposure given that both have alternative sources of income (see Appendix 11). As highlighted, Sudeep owns an auto-rickshaw, a lucrative income source given that it is the only one in the village, and Ashna's husband works in construction. Both were debt-free prior to the 2010/2011 season. Ashna, however, was obliged to take out a loan during the 2010/2011 season for her daughter's hospital care, as well as to pay labour costs for cultivation. She was unable to repay this loan as her entire crop was lost due to the flooding.

As Table 6.4 highlights, participants noted a number of sources of borrowing. In Bantala, Chitta (male, dominant caste, semi-medium cotton land-holder) is a director of a co-operative bank in the village. This bank lends only to land-holders and solely for the purposes of farming. In Nandanapuram, both Scheduled Caste female participants (Salma who is landless, and Ambu who is a cotton small-holder) are members of Self-Help Groups which facilitate borrowing for any purpose. The local NGO, Crops Jangaon, with which Orgampalle is associated, runs a savings society which, at the time of my interview with its Director (30/8/2010), had 12,000 members. Participants paid Rs 20 per month into the scheme and could access loans for any purpose. All of these, in conjunction with loans from family and members of one's own caste, represented low-interest, low-risk sources of borrowing.

As highlighted, however, borrowing from Micro-Finance Institutions is particularly high-risk, not only given their high interest rates but also their coercive methods of extracting payments. In Bantala, Ashna (female, SC Madiga, marginal cotton-holder) borrowed from an MFI and notes the risks associated with default. 'First, you lose your respect in the village,' she observes. 'Second, they will hand you over to the police.' The sequence in which these fears are recounted highlights the significance of shame in relation to failure to repay debt in the villages. Ashna has witnessed representatives from an MFI—two men on a motorbike—coming to the village to collect loan repayments. Her facial expression, as she recalled this, revealed the sense of dread which their visit represented for her.

Appendix 14 provides details of the main reasons given for debt among small-holders in Bantala. These include dowry, house construction and costs for hospital care, as well as cultivation costs (seeds, inputs and labour). The two female landless participants in Bantala have secured loans from landlords in the village to pay for hospital care and for a training scheme launched as part of a Rural Poverty Reduction Project. They are now both obliged to work in semi-bonded conditions (they are permitted to work for others, but must prioritise their creditor when he requires them). Loan repayments will be deducted from their daily rate by their creditor. Given that they are either illiterate (Nipa) or semi-literate (Abani), these participants are reliant on the land owner to tell them when the debt is paid.

In Nandanapuram, it is again the Forward Caste (Rs 250,000) and medium Bt cotton cultivators (Rs 236,667) who are most indebted (see Table 6.4). They are closely followed by marginal cultivators in the village (Rs 215,000). Reasons given include cultivation and labour costs, education of children and, for Rashi (female, FC Brahmin, marginal cotton cultivator), health care and house construction (Rs 500,000 is being spent on a sizable house to be shared in a joint family arrangement with her sister). This suggests that the legitimation of indebtedness by Forward Castes forms part of their quest to modernise in keeping with their upper caste expectations. In Rashi's case, even the vulnerability associated with her widow status and gender does not eliminate this Forward Caste tendency towards conspicuous consumption funded by debt.

The difference between the debt exposure of Bt cotton and organic cultivators provides some indication of the impact of the extra costs of Bt cultivation in addition to the other reasons for indebtedness prevalent in the Telangana context. The average expenditure on pesticides for Bt cotton cultivators in the study was Rs 3500 per acre, with a further Rs 2400 per acre for fertilisers and an average Rs 840 per acre on Bt cotton seeds (as opposed to Rs 440 per acre for non-Bt seeds) (Desmond 2014: 187–188).[6] This meant an average cost for seeds, pesticides and fertilisers of Rs 6740 for Bt cotton cultivators.

Participants in Orgampalle incur neither pesticide nor fertiliser costs. Their main cost is labour and, in the rare cases where oxen are not owned (only Amita: female, BC Chakali, marginal cotton-holder, does not own oxen), the charge associated with leasing these. As can be seen from Table 6.4, the main reasons for indebtedness in Orgampalle include dowry, labour costs, house construction and hospital charges. Although small and marginal cultivators in Orgampalle are still in debt due to reasons other than cultivation costs (Rs 50,000 and Rs 47,250, respectively), they are significantly less exposed to debt than Bt cotton cultivators in these categories. In Orgampalle, Pradnesh (male, dominant caste, medium land-holder), the influential village elder, is debt-free.

The two NPM participants in Nandanapuram (Nand: male, BC Yadava, small-holder of cotton; and Nishok: male, SC Madiga, marginal cotton cultivator) were both debt-free prior to the 2010/2011 season. However, Nishok was obliged to access a loan (Rs 25,000) as a result of labour charges during the research period. Nand has bought land, constructed a house and cleared debts inherited from his father, and is now debt-free. He does, however, also receive a salary as an NGO employee (see Appendix 12).

As Table 6.4 highlights, females are more indebted than males in both Nandanapuram (Rs 156,667) and Orgampalle (Rs 75,000). In Nandanapuram, the three female participants are widows—Rashi (FC Brahmin, marginal cotton-holder); Ambu (SC Madiga, cotton small-holder) and Salma (SC Madiga, landless). The two female participants in the village who cultivate Bt cotton have incurred debt due to house con-struction, dowry and health care, as well as input and labour costs (see Appendix 15). The average debt for females in Nandanapuram also includes that of Salma (female, SC Madiga, landless) who accessed a loan through a Self-Help Group to lease land on which she suffered crop loss (paddy). She now owes Rs 10,000.

Along with indebtedness, the risk of suicide remains a very real aspect of village life. In Nandanapuram, Salma (female, landless, SC Madiga) was on her way to the local hospital following an interview with us. She was visiting a cousin who had tried to commit suicide by ingesting pesti-cide following an argument with her husband about a loan she had taken out.

In Orgampalle, one of the cultivating females, Amita (a BC Chakali marginal cotton cultivator) is a widow; the other, Aruni (dominant caste, marginal cotton cultivator), has a husband who is a herder. Both females have incurred debt due to dowry, as well as a (now failed) bore-well in the case of Aruni and healthcare in the case of Amita (see Appendix 16). Despite the fact that, as Table 6.4 highlights, the debt levels for females in Orgampalle are higher than males, they are signifi-cantly lower than those incurred by the Bt cotton cultivating females in Nandanapuram.

A number of participants noted the government's debt waiver initia-tive.[7] This was launched by the government in 2008 in an attempt to ease the situation of indebtedness in the rural areas. The waiver was offered only for loans taken out in banks and co-operative credit institutions for direct agricultural purposes. The conditions of the waiver were a source of confusion in the villages, particularly in the case of the most vulnerable.

Nipa (female, SC Madiga, landless) in Bantala, for instance, had hoped her loan of Rs 24,000 would be waived. The borrowing was associated with her surgery in a private hospital and training undertaken by her husband to become a cobbler as part of an Andhra Pradesh Rural Poverty Reduction Project. (She showed me the certificate he had obtained for this). She claimed 'YSR [Y. S. Rajasekhara Reddy, the Chief Minister who introduced the loan waiver] cancelled others' loans so I was hopeful mine would be cancelled too. But it was not.'[8]

Natesh (male, BC Gowda, cotton small-holder) who had contemplated suicide due to indebtedness had also hoped his loans would be waived as part of debt relief but, because his debts largely arose due to dowry payments not covered under the scheme, they were also not waived. Despite the sale of 6 acres of his land a number of years ago to ease his debt exposure, Natesh is again Rs 65,000 in debt. During the 2010/2011 season, he was obliged to sell his oxen in order to pay for labour costs.

Aruni (female, dominant caste, marginal cotton holder) in Orgampalle, however, noted that she had had Rs 3000 of her bank loan cancelled. The borrowing had been taken in part payment for a borewell which has long since failed. The remainder of the debt was sourced from an MFI so was not covered by the waiver.

As can be seen from Table 6.4, despite this loan waiver for agricultural loans from formal sources in 2008, average debt levels across all categories of land-holding remain high, particularly for Bt cotton cultivators. The sources of debt indicate that the loan waiver, which focussed on loans taken from formal sources for agricultural purposes, would only ever have dealt with a portion of the indebtedness of participants given the multiple sources of loans and the non-agricultural, as well as agricultural, reasons for borrowing.

Participants, particularly the more vulnerable, highlighted their reliance on the National Rural Employment Guarantee Scheme (NREGS), particularly for the 3 months from March to May after the agricultural season has ended. In Bantala, for instance, Nipa (female, SC Madiga, landless) notes, 'this rural scheme is there. During that time, construction work will be available, or work grinding gram.'[9] With this work, she can keep up the repayments for the loans she has taken from a bank and a village land owner.

The role of middlemen in the negotiation of risk in the villages is also significant. As noted, Bt cotton was first introduced in Bantala and Nandanapuram following the persuasion of key villagers by seed dealers that they should try it. The restricted supply of non-Bt seeds was highlighted in a personal interview (22/11/2010) with a seed dealer in Warangal. The respondent claimed that he had stopped supplying non-Bt seeds due to reduced demand. He also noted, however, that his commission for the sale of a packet of Bt seeds was Rs 80–100, depending upon the size of the order. For non-Bt seeds, it was just Rs 50.[10] This means that, without the co-ordination of demand for non-Bt cotton seeds across the villages which NGOs organise, there would be no alternative other than Bt cotton given the minimum order quantities for non-Bt seed supply.

The impact of middle men on output prices was also noted during a trip to a cotton market in the Warangal district. This was undertaken with Prakash (male, BC Yadava, marginal cotton-holder) from Orgampalle (field note extract, 26/11/2010). Here, the power of the trader was very much in evidence. The hessian sacks in which the cotton was transported by auto-rickshaw to the market were slit with a slash hook in order to check the lustre and staple length of the cotton which would determine the price. Once the sack was slit, however, no other trader would offer a price on that cotton. Hence, the farmer was obliged to negotiate with the original trader, rather than being free to seek a better price elsewhere.

Given the catastrophic 2010/2011 season and the reports of farmer suicides, the government vowed to compensate cultivators for their crop loss *(Deccan Chronicle, 27/12/2010)*. It was argued however that, while the government of Tamil Nadu offered Rs 4500 per acre, the government of pre-secession Andhra Pradesh government proposed only Rs 2400 per acre. Given the average cultivation cost per acre for marginal and small cultivators in the current study was between Rs 15,000 and Rs 21,000 respectively, it is clear that the suggested compensation was inadequate. As at March, 2011, no compensation had been received by the cultivators in the study. Likewise, participants claimed not to have taken out crop insurance due to the costs involved.

This chapter has explored the way in which powerful villagers are influential in both the legitimation and delegitimation of Bt cotton. It has also illustrated the impact of the village power structure on the negotiation of risk. Bantala, a village in which asset ownership is highly concentrated and Bt cotton strongly legitimated due to its economic benefits, has the highest average debt exposure of the study.

Average debt levels among Bt cotton cultivators in Nandanapuram, a village characterised by its contested power structure, are also significant. The absence of an over-riding authority in Nandanapuram has led to a mixed approach to risk negotiation. In Orgampalle, the influence of a debt-free charismatic elder and the absence of aspirational Forward

Castes have contributed to a more collective approach to risk negotiation. This has resulted in the average debt exposure of Orgampalle participants being the lowest of the study.

The analysis of the debt exposure of participants undertaken here supports the view that Bt cotton is by no means the sole reason for participant indebtedness. However, such an assertion does not provide the full picture either. Instead, the additional debt exposure of Bt cotton cultivators suggests the contribution of the specific costs associated with Bt cotton—the seeds, pesticides and fertilisers—to the overall indebtedness of participants.

The analysis highlights that indebtedness is also exacerbated by the conspicuous consumption of high status Bt cotton cultivators and the emulation of their cultivation practice and lifestyles by lower castes. In Bantala, this consumption is funded through debt, a factor which contributes to the diffusion of indebtedness, as well as Bt cotton, in the village. In Nandanapuram, the indebtedness of participants itself provides the main basis for the ongoing adoption of Bt cotton in an attempt to secure higher yields, even while Bt cotton cultivators in the village strongly delegitimate their own farming practice (see Fig. 6.1).

The dynamic process through which attempts are made to gain recognition, representation and resources to address risk as a political concern, as well as the perceived role of the state in risk negotiation, relates to the legitimation of democracy in the villages. This will now be explored.

Notes

1. Durga is the mother goddess of the Hindu pantheon. She is purported to protect against evil demons that threaten peace and prosperity. Ganesh, the elephant-headed god, is the patron of arts and the sciences. Both are associated with popular festivals in Warangal—*Durga Puja* (September/October) and *Ganesh Chaturthi* (August/September).
2. The official minimum wage rate in pre-secession Andhra Pradesh for the 2010/2011 season was Rs 112 for agricultural labour (Kolamkar 2010: 30). It is clear that wage rates have risen significantly in just over a

decade—the minimum rate was Rs 35 per day in 1996. This pressure on wage rates, however, must be borne by cultivators. Although the official minimum labour rate does not differentiate between male and female labourers, the gender difference in pay rates is legitimated by villagers on the basis of the division of labour. Men's work includes the spraying of pesticides and the ploughing of land with oxen. This is considered heavy work, and so is better paid, despite the strenuous, back-breaking nature of the work associated with sowing, weeding and harvesting which is performed by females.

3. There is evidence of migration in all three villages. In Bantala, I was offered the use of a house in the dominant caste ward whose owner had migrated to Mumbai. Two Forward Caste participants in both Bantala (Pavan) and Nandanapuram (Nikhil) had children studying in Australia and the United States, respectively, while the son of Achanda (male, BC Chakali, cotton small-holder) in Orgampalle lives in a nearby town. Ashna (female, SC Madiga, marginal cotton-holder) migrated to the village after her marriage, but both she and her husband had spent a number of years living in Mumbai.

4. A quintal, the unit of measure for cotton yields, is equivalent to 100 kgs.

5. In an interview with a cultivator in Nandanapuram which is not included in the analysis, he asserted that dry, discoloured patches on his feet were due to his walking on Bt cotton fields. None of the other cultivators or wage labourers reported any problems.

6. The price of Bt cotton seeds varied depending upon their source and ranged between Rs 750 and Rs 1250 per 450 gm pack sufficient for one acre. A number of participants claimed that larger farmers would buy up the stocks of Bt cotton seeds at traders and sell them to smaller farmers at inflated prices. The cost of non-Bt cotton seeds ranged from Rs 450 (the most common price) to Rs 600, while Nand, the NPM farmer, obtained non-Bt seeds for free given his NGO employment.

7. This refers to the Agricultural Debt Waiver and Debt Relief Scheme (ADWDRS). Details are available at: http://pib.nic.in/newsite/PrintRelease. aspx?relid=104122. Accessed on 30/3/2017.

8. The Chief Minister of Andhra Pradesh at that time, Y.S. Rajasekhara Reddy, was instrumental in persuading the Central Government to announce the loan waiver. Available at: http://www.thehindu.com/ todays-paper/tp-national/tp-andhrapradesh/YSR-credited-with-loan-waiver-scheme/article15262124.ece Accessed on 30/3/2017.

9. Ground red gram, or pigeon pea, is used to make *dal.*
10. This absence of non-Bt seeds in the market is highlighted in a 2011 documentary entitled *Bitter Seeds* produced by Micha Peled. The film explores the experience of Bt cotton farmers in Maharashtra. Details are available at: http://www.itvs.org/films/bitter-seeds Accessed on 30/3/2017.

Bibliography

Beck, U. (1992). *Risk society: Towards a new modernity.* London: Sage.

Desmond, E. (2014). *The legitimation of risk and democracy: A case study of Bt cotton in Andhra Pradesh, India.* Cork: University College Cork. Available at: https://cora.ucc.ie/handle/10468/1688/

Goyal, L. C. (2015). *Accidental deaths and suicides in India, 2014.* Report from National Crime Records Bureau, New Delhi.

Henrich, J. (2001). Cultural transmission and the diffusion of innovations: Adoption dynamics indicate that biased cultural transmission is the predominate force in behavioural change. *American Anthropologist, 103*(4), 992–1013.

Kolamkar, D. S. (2010). Report on the working of the Minimum Wages Act, 1948, for the year 2010. Government of India Ministry of Labour and Employment, Labour Bureau. http://labourbureau.nic.in/REP_MW_2010.pdf. Accessed 22 July 2013.

Mishra, S. (2007). *Risks, farmers' suicides and agrarian crisis in India: Is there a way out?* Mumbai: Indira Gandhi Institute of Development Research.

Omvedt, G. (1994). *Dalits and the democratic revolution: Dr. Ambedkar and the Dalit movement in colonial India.* London: Sage.

Revathi, E. (2009). 'Farmers' suicides in Andhra Pradesh: Issues and policy concerns. In S. M. Dev, C. Ravi, & M. Venkatanarayana (Eds.), *Human development in Andhra Pradesh: Experiences, issues and challenges.* Hyderabad: Centre of Economic and Social Studies.

Stone, G. D. (2007). Agricultural deskilling and the spread of genetically modified cotton in Warangal. *Current Anthropology, 48*(1), 67–103.

Townsend, R. M. (1994). Risk and insurance in village India. *Econometrica, 62*(3), 539–591.

Vasavi, A. R. (2012). *Shadow space: Suicides and the predicament of rural India.* New Delhi: Three Essays Collective.

Walker, T. S., & Ryan, J. G. (1990). *Village and household economies in India's semi-arid tropics.* London: The Johns Hopkins University Press.

7

Analysis II: Democracy and State Legitimacy in the Villages

This chapter explores the impact of power relations on the legitimation of democracy in the villages. It examines the way in which the legitimation of democracy is mediated through village power structures with respect to three aspects: (a) institutionalised democratic practice associated with the electoral process and *gram sabha* meeting attendance, (b) the demands for a separate state of Telangana and (c) the intersection of the legitimation and delegitimation of Bt cotton with the wider struggle for legitimacy associated with the pre-secession state government.

Voting and Gram Sabha Attendance

All participants in the study claimed to vote in local elections. This legitimation of voting is strongly linked to attempts to secure access to resources as the means to alleviating exposure to risk. This is asserted even by the more powerful participants. In Bantala, Sudhakar (male, FC Reddy, semi-medium land-holder) argues, 'if I cast my vote, I may get benefit.' Likewise, in Nandanapuram, Charan (male, FC Brahmin, semi-medium cotton cultivator) claims, 'the person who gets elected may be of some use.'

© The Author(s) 2018
E.L. Desmond, *Legitimation in a World at Risk*,
https://doi.org/10.1007/978-981-10-6065-6_7

A number of participants assert their view of voting as a social norm. Nikhil (male, FC Reddy, semi-medium land-holder) in Nandanapuram claims, 'we are under a compulsion to vote....Otherwise, we wouldn't.' And Ashna (female, SC Madiga, small-holder) in Bantala states, 'if they say to cast my vote, I'll go and cast it.'

For others, voting is associated with a discourse of rights. Pradnesh (male, dominant caste, medium land-holder), the village elder in Orgampalle, argues, 'only if I vote, will I be able to ask for my rights.' In Bantala, Nipa (female, SC Madiga, landless) also asserts, 'because I have the right to vote, I vote.' And Sudeep (male, SC Mala, marginal land-holder), in Bantala, illustrates well the centrality of the right to justification within the legitimation of risk and of democracy when he argues:

> since we are living in a democratic country, this has given us the right to ask for our own rights and to fight for rights. Now I have enough freedom to go and ask what rate I'm getting [for my cotton], and why I'm getting it. I can fight for my own sustenance and rights. No-one can stop me from asking for my rights.

Even the landless participants in all three villages assert their right to vote and its compulsory aspect. Sajan (male, dominant caste, landless) in Orgampalle asserts, 'I have the right to vote, so I exercise that right.' And Ranjan (male, BC Chakali, landless) in Nandanapuram claims, 'if I didn't vote, somebody would come and make me.'

There is also evidence of the 'money politics' described by Powis (2003: 2620). In Nandanapuram, Ranjan (male, BC Chakali, landless) notes, 'they [politicians] offer 50, 60 rupees for my vote...Every party bribes, usually with alcohol. I don't drink, so they give me money.' And Charan (male, FC Brahmin, semi-medium cotton cultivator) in Nandanapuram claims, 'they bribe me for my vote. I don't take any bribe, but I vote.'

Villagers also indicate signs of a conformist bias (Henrich 2001: 997) with regard to voting behaviour. This is asserted by Amita (female, BC Chakali, small-holder) in Orgampalle who claims, 'everyone votes, so I also go to vote.' She also notes the potential opportunities to enhance the ability to negotiate risk associated with democratic practice, stating 'maybe they [elected politicians] might undertake some welfare measures for the sake of the village.'

According to the system of reservations, the *sarpanch* seat had been reserved for Backward Caste females in Nandanapuram and Bantala in the panchayat elections 4 years earlier. For those elections, the seat was an open one in Orgampalle.[1] Participants noted how, even with reservations, *sarpanch* elections are strongly mediated by the village composition and associated power structure.

In Nandanapuram, Anshul (male, BC Gowda, marginal cotton-holder) notes the strong influence which FC Reddys continue to exert over the process, even with reservations, through their land ownership. He claims, '[i]f a Reddy owns 20 acres of land and 50 labourers from the *Dalit* community work in his fields and he tells them that we have nominated a person from a certain community and you should vote for them, they will oblige.'

The mediation of reservations of PRI positions through the power structure was also evident in Bantala given that, although a female from a Backward Caste had been elected, she was from the dominant caste in the village. It was also her husband who, to all intents and purposes, was the *sarpanch*. This was evident from the fact that all participants used the male pronoun to refer to the *sarpanch*.

The intersection of the institutionalised PRI structure with that of the traditional power structure associated with village elders was also apparent, particularly in Orgampalle. The *sarpanch* for Orgampalle, although based in another village, is generally well respected.[2] Nonetheless, all participants refer to Pradnesh, the village elder, as the most influential person in Orgampalle.

Nirmal (male, BC Yadava, cotton small-holder), for instance, claims, 'before going to the *sarpanch* [for help with a problem], I would meet with [Pradnesh]. If you go directly to the *sarpanch*, you may have to make a complaint at a police station. That's why I'd go to [Pradnesh].' This highlights that the *sarpanch* position is viewed as linking the village with more official structures of governance beyond the village. Sajan (male, dominant caste, landless), the most vulnerable participant in Orgampalle, seems unaware of what the PRI structure entails in terms of the mitigation of his risk exposure. He claims, 'I don't know what he [the *sarpanch*] is doing [i.e. what he is responsible for]. I have no idea.' Thus, the participant who is most in need of the *sarpanch*'s assistance is unaware of what the official's duties are in relation to his risk alleviation.

The village assembly or *gram sabha* meeting is largely regarded as the cornerstone of decentralised democratic practice in India. According to the Andhra Pradesh Panchayat Raj Act (1994), village assemblies are legally obliged to convene at least twice per year.[3] Villagers are offered no incentive to attend, other than to have the chance to contribute to the collective decision-making of the village, hold village officials accountable and obtain information on development programmes and government initiatives.

Details of *gram sabha* meeting attendance in the villages by the land-holding, caste and gender of participants are provided in Table 7.1.

As can be seen from Table 7.1, attendance at the *gram sabha* is far less strongly legitimated than voting. Just over one-third of participants in Bantala claim to attend the *gram sabha* meeting and 20 per cent of

Table 7.1 *Gram sabha* meeting attendance by land-holding, caste and gender

	BT	NP (Bt)	NP (NPM)	OR
Participant numbers	N = 8	N = 8	N = 2	N = 8
Average attendance (% of participants)	**37.5**	**10**	**10**	**62.5**
Attendance level % by total land-holding[a]				
Marginal	0	0	–	12.5
(0.1–2.5 acres)				
Small (2.6–5 acres)	12.5	10	0	**37.5**
Semi-medium (5.1–10 acres)	25	0	10	–
Medium (10.1–20 acres)	0	0	–	12.5
Landless	0	10	–	0
Attendance level % by caste				
FC	0	0	–	–
DC	12.5	–	–	25
BC	12.5	10	10	**37.5**
SC	12.5	0	–	–
Attendance level % by gender				
Male	25	10	10	**62.5**
Female	12.5	10	–	0

BT Bantala, *NP (Bt)* Nandanapuram Bt cultivators, *NP (NPM)* Nandanapuram NPM cultivators; *FC* Forward Caste, *DC* Dominant Caste, *BC* Backward Caste, *SC* Scheduled Caste

[a]Table 6.4 on debt levels in the previous chapter focussed on cotton land-holding given that the objective was to examine the impact of Bt cotton cultivation on debt levels. However, because the focus in Table 7.1 is on the impact of the relative power of participants on *gram sabha* attendance, their total land-holding is used as the basis for comparison

participants in Nandanapuram (one Bt cotton and one NPM cultivator). The highest attendance rate is in Orgampalle, where just over 62 per cent of participants stated that they attended the meeting.

Of the three participants who claimed to attend the meeting in Bantala, two were male—Chitta (dominant caste, semi-medium land-holder) and Natesh (BC Gowda, semi-medium land-holder), a village elder. Ashna (SC Madiga, small-holder) was the only female participant in the village who claimed to attend the meeting.

Of the three participants who stated that they attended the meeting in Nandanapuram, two were male: Nand (male, BC Yadava, semi-medium land-holder) and Anshul (male, BC Gowda, small-holder). The only landless participant to claim attendance was Salma (female, SC Madiga). Salma's (now deceased) husband was a former *sarpanch* of the village, so she is very politically aware. In Orgampalle, all five attendees were male.

As Table 7.1 highlights, Forward Caste participants do not attend the meeting in either Bantala or Nandanapuram. (Orgampalle is comprised solely of Backward Castes). In Bantala, Pavan (male, FC Vaishya, medium land-holder) argues, '*gram sabha* meetings come under politics, so since I'm a farmer, I have to go to my work.' And, in Nandanapuram, Rashi (female, FC Brahmin, small-holder) argues, 'because of my religion and my caste, we strictly follow purity, so we don't engage in political activities.' Her status as a Brahmin did not however, as will be seen, prevent her from having strong opinions on the perceived need for a separate state of Telangana.

As highlighted, while Orgampalle had the highest average attendance rate, all attendees were male. This was due to a perception that an interest in politics was not appropriate for females. Aruni (female, dominant caste, small-holder) claims, 'women like me never attend [*gram sabha*] meetings.' For Aruni, the responsibility for managing political affairs and monitoring the performance of the *sarpanch* has been passed to village elders. She claims, 'village elders know if he [the *sarpanch*] is doing a good job or not. I don't know. That is not my level. That is the level of elders.' Likewise, Amita (female, BC Chakali, small-holder) asserts, 'I don't go to meetings. I just look after my housework.'

Cultivators in Nandanapuram assert their lack of time as the main reason for their non-attendance. Nishok (male, SC Madiga, small-holder) claims, 'if I keep attending these meetings, I'll not have much time to attend to my fields.' And Ambu (female, SC Madiga, small-holder) argues, 'I can't afford the time to attend [*gram sabha*] meetings... I go to women's [Self-Help Group] meetings.' The statement highlights that, in Ambu's case, Self-Help Group attendance is prioritised over the *Gram Sabha* given that the former is seen as more relevant to the risk negotiation associated with her gender.

Apart from Salma (female, SC Madiga) in Nandanapuram, the landless participants who could most benefit from patronage opportunities do not attend the meeting. Ranjan (male, BC Chakali) in Nandanapuram states, 'Earlier I used to go to *gram sabha* meetings, but I never understood anything. I used to just stand there.' And Abani (female, BC Padmashali) in Bantala claims, 'I don't have time. I have to get the children to school, and cook.' In Orgampalle, Sajan (male, dominant caste) argues, 'Why should I go? I'm not well. I would have to pay to attend.' (The Orgampalle *gram sabha* is held in the neighbouring village which Sajan would require transport to reach because of his age and poor mobility).

As in the case of the legitimation of risk explored in the previous chapter, the village power structure has a significant impact on the way the practice of democracy is legitimated in the villages. In Nandanapuram, Anshul (male, BC Gowda, small-holder) argues, 'in politics, the word of the wealthy prevails, especially the upper castes, the Reddys. Under their rule, other castes have to live their lives.'

Anshul (male, BC Gowda, small-holder) notes the numerical strength of SC Madigas in Nandanapuarm; he highlights however that, because SC Madigas are obliged to work for Reddy landowners, their voting behaviour is strongly influenced by the Reddys. He claims, 'the deciding factor is votes from the SC community as they are a majority in the village. But the Reddys tell them who to vote for.' (This is in contrast to Robinson's (1988) finding in her study that the abolition of bonded labour had made agricultural labour relatively politically autonomous, particularly with regard to the dominance of the Reddys).

Villagers in Nandanapuram also describe the way in which patronage opportunities intersect with village power relations and party affiliations. Nand (male, BC Yadava, semi-medium land-holder) argues, 'the *panchayat* is supposed to select the poorest of the poor for the policies, but the *pan-chayat* or the village elders select their favourites.' Nand also notes the influence of party politics on the ability to access resources, claiming 'the *sarpanch* favours his *[sic]* own party members.' Anshul (male, BC Gowda, small-holder) highlights the intersection of caste with the official power structure, arguing 'if the *sarpanch* is from a particular [caste] community, the community gets benefited in government welfare schemes.'

There is also a certain legitimation of corruption in the villages, particularly among the most vulnerable. This suggests the ongoing influence of Telangana's imperial past, and its association with privileged rulers responsible for the dispensing of patronage, on contemporary democratic practice. Thus, Amita (female, BC Chakali, small-holder) in Orgampalle argues, 'they [politicians] are important people so if they manage to get money they may keep some for their own purposes and what we get is what we get.'

The idea of the *gram sabha* as a place for deliberation on village problems is noted by just one Orgampalle participant. Akhil (male, dominant caste, small-holder) claims, 'we talk about the problems in the village in the *gram sabha* meetings.' The greater caste homogeneity and limited caste conflict in Orgampalle no doubt support such a deliberative component to risk negotiation in the village. This deliberation clearly remains, however, male dominated.

The sole female participant in Bantala who claimed to attend the meeting, Ashna (SC Madiga, small-holder), was not present at the *gram sabha* which I attended in the village (04/10/2010).[4] In fact, the only participant present on this occasion was Natesh (male, BC Gowda, semi-medium land-holder). Villagers had been summoned to the meeting by the banging of a drum in all caste wards by the elderly SC Madiga village servant, Neelam, at the behest of Pallav.

The meeting was held outdoors beneath the trees in the courtyard of the *panchayat* office, a number of benches having been arranged around a central table by Neelam. My translator noted that, prior to the meeting, Pallav was himself aggressively insisting with villagers that they attend. His wife (the elected *sarpanch*) and a representative from the Mandal were seated at the table (joined, on this occasion, by myself and my trans-lator, at Pallav's insistence).

During the meeting, 25 males assembled on the benches. Five females stood silently in the background. Although not seated at the head table, Pallav directed the discussion from the benches, and it was to him that all participants addressed themselves, his wife being largely ignored.

Discussion at the meeting revolved around the need to remove the names of deceased villagers from the pension list, as well as the requirement for new drains, road widening and a water tank in the village. There was also heated argument from some villagers who claimed that plans to widen a road would cause structural damage to their homes (translator's notes: 04/10/2010). Surprisingly, Pallav had invited journalists from the local press who waited for me after the meeting and questioned me as to my experience of Indian democracy. This left me with an unconfirmed, but distinct, impression that the meeting and the press attention had been orchestrated by Pallav because of my presence in the village - perhaps to gain local prestige - though it was impossible to be sure of his motives in notifying the press in this way.

In a conversation with Sudeep (male, SC Mala, marginal landholder) in his auto-rickshaw following the meeting, he claimed that the *gram sabha* was a waste of time because the people who were most in need of assistance could not attend due to their need to work in the fields. He also argued that villagers who were entitled to Indiramma houses were not getting them and highlighted a dispute between the *sarpanch* (by whom he meant Pallav) and the village representative at the mandal level. This, he claimed, meant that the names of many villagers had been removed from the housing list (field note extract: 04/10/2010).

The centrality of the *sarpanch* in the accessing of resources to negotiate risk through the democratic process was also highlighted by Ranjan (male, BC Chakali, landless) in Nandanapuram. Ranjan claimed to have stopped receiving his pension as a result of a dispute with the *sarpanch's* wife (who was actually the *sarpanch*). He claimed that all the SC Madigas in the village asserted his cause to the *sarpanch*, but to no avail. This meant that he and his ill wife were obliged to live off the meagre income from his laundry work and to collect the rice used in festivals to supplement their PDS allowance.

The Demand for Secession

Demands for a separate state were reaching a crescendo during the research period. Roads to the villages were frequently blocked by *rokos* (blockades), with protestors stopping traffic to sing songs, make speeches and wave flags. Towns were alive with colourful bunting, and music and speeches blared from loud speakers to the assembled crowds.

In the rural areas, hunger strikers protesting for a separate Telangana would sit beneath multi-coloured pagodas, garlands of jasmine hung around their necks to indicate their fast. In Bantala, relay hunger strikes were orchestrated by Pallav. Villagers took turns in their 'fasts' which involved eating breakfast, missing lunch and taking an evening meal. Males and females fasted on alternate days. I spent time in one such fast with a number of Bantala males, including Pallav. The protest was an exhilarating, lively social occasion, with songs and speeches in Telugu resonating throughout the village from loud speakers. The vibrancy, colour and lively excitement associated with Indian democratic practice is part of its particular 'vernacularisation' (Michelutti 2007).

Remnants of the historic Telangana armed struggle were also evident in the villages. A red flag bearing the hammer and sickle of the CPI (Marxists) fluttered aloft a flagpole in the centre of Bantala, and the hammer and sickle emblem was visible in graffiti throughout the village. In Nandanapuram, memories of the struggle were still very much alive. I was taken by Nand (male, BC Yadava, semi-medium land-holder) to a monument erected in honour of those who had died fighting to free Hyderabad State from the control of the Nizam. Meanwhile, Charan (male, FC Brahmin, semi-medium land-holder) was an elderly Brahmin for whom the struggle had formed the backdrop of his teenage years. He recounted stories of the flight of the local landlords during the conflict. He told how the village was caught in the crossfire of the struggle, plundered by the Nizam's *Razakars* during the day and by the communists and Congressmen at night.

Ranjan (male, BC Chakali, landless) also spoke of the historical impact of the Naxalites in Nandanapuram. He claims, 'Naxalites used to be very active here. They kept Reddys in check. When the bonded labour was beaten by the upper castes, they would complain to the Naxalites who would beat the upper castes. Without them, we would still be bonded labour.' It is interesting to note that it is to Naxalite activity, rather than to state legislation, which Ranjan attributes the releasing of agricultural labourers from their situation of bondage.

Given the generally energised situation at the time of the research, most participants were keen to talk with me on the theme of Telangana. Their animation on the subject of a separate state permitted insights into the normative perspectives associated with the legitimation of democracy in the Telangana context.

The vast majority of villagers argued that a separate state was necessary in order to secure a more just allocation of resources. This was associated with the perspective that powerful politicians and business people from Coastal Andhra were diverting resources from the region and that a smaller state governed by Telangana politicians would allow Telangana inhabitants greater access to, and control over, their own resources. This view was asserted across the villages, regardless of the land-holding, caste and gender of participants. In Nandanapuram, Anshul (male, BC Gowda, semi-medium land-holder) argues, '[a separate state of] Telangana would be advantageous in terms of access to natural resources. It would benefit all castes equally if Telangana came.'

For many of the landless Scheduled Caste participants, secession represented hope, not necessarily for themselves but for their children. In Bantala, Abani (female, BC Padmashali, landless) claims, 'I will not receive anything if we get Telangana, or if we don't. But my children may get the benefit.' Likewise, Salma (female, SC Madiga, landless) in Nandanapuram asserts, 'I believe Telangana will bring jobs for my children. Our generation has lost the benefits. We're fighting for the next generation.'

The delay in accessing the longed-for benefits of secession at times skipped a number of generations. So, Natesh (male, BC Gowda, semi-medium land-holder) in Bantala argues, 'if we get Telangana, my sons may not get a job, but my son's sons may get one.' This suggests that the new state was seen by participants as a new beginning for the young.

The view relates to the important insight from the Srikrishna Report (Srikrishna et al. 2010: 162–163) that many of Telangana's youth were the first generation of college graduates in their families. The opportunities which it was hoped such an education would bring were particularly important for those from the lower castes given that the previous generation had often been denied such chances as a result of the poverty and exploitation associated with the Nizam's rule.

The theme of employment was closely connected to that of education and was central for all participants. In Nandanapuram, Anshul (male, BC Gowda, small-holder) asserts, 'if Telangana comes, there will be abundant opportunities…Like employment for the educated…and sufficient agricultural employment for the landless.'

The much sought-after government jobs were seen as particularly crucial. Pradnesh (male, dominant caste, medium land-holder) in Orgampalle observes, '[t]here won't be much change for farmers, but a separate state would definitely be helpful for employment in government. If there are Telangana people employed there, it would be helpful for the region.'

It was felt that those in Coastal Andhra were to blame for the lack of employment opportunities for Telangana's youth. Rashi (female, FC Brahmin, small-holder) in Nandanapuram argues, 'more than half the employment opportunities are taken by those from [Coastal] Andhra, while our graduates have no employment.' Likewise, Charan (male, FC Brahmin, semi-medium land-holder) claims, 'our children are getting educated, but there is nowhere for them to work because those from [Coastal] Andhra have taken the employment opportunities.'

As highlighted, a central issue which emerged in relation to the demands for a separate state was the sense of injustice at the unequal allocation of resources between the regions of the pre-secession state, particularly with regard to Coastal Andhra. The anger at Telangana's discrimination in this regard was asserted by all villagers, again regardless of their land-holding, caste or gender. In Nandanapuram, Rashi (female, FC Brahmin, small-holder) asserts, 'no development projects are coming to us. Everything is going to the [Coastal] Andhra region. Though we have the Nagarjuna Sagar [dam] located here, more than half the water goes to [Coastal] Andhra.'[5]

The concern with water was prominent among all land-holders, regardless of caste and gender. Anshul (male, BC Gowda, small-holder) in Nandanapuram argues, 'it would definitely be advantageous to us if there is a separate Telangana in terms of access to natural resources....If Telangana comes, I will be able to get more resources like water.'

In Bantala, Pavan (male, FC Vaishya, medium land-holder) claims,

> Now that both [Coastal] Andhra and Telangana are combined, [Coastal] Andhra people are using all our resources like mines, water and jobs. If we can get a separate Telangana, we can use our own resources for our own benefit.

Securing access to the region's resources was seen as vital to allowing cultivators in Telangana to negotiate agrarian risk. In Orgampalle, Nirmal (male, BC Yadava, small-holder) argues, 'people from [Coastal] Andhra are coming to Telangana, and they're developing. We're not developing. If Telangana is separated, we will develop.'

It was also hoped that the new state would lead to improvements in the provision of public services. Nirmal (male, BC Yadava, small-holder) in Orgampalle observes, 'we're paying very high prices to private hospitals. If we get Telangana, we'll benefit from government [state-funded] hospitals.' And Salma (female, SC Madiga, landless) in Nandanapuram argues, 'power stations are based in Telangana, but half of the electricity goes to the [Coastal] Andhra region.'

There was also a demand for recognition of the cultural identity linked to the shared history and sense of place associated with Telangana. This was asserted through the frequent use of the Telugu word *desam*, to signify land or country, in response to the question: 'What does Telangana mean to you?' In Bantala, Chitta (male, dominant caste, semi-medium land-holder) responds, '[i]t's our country [*desam*], and we should have that.' And Sudeep (male, SC Mala, marginal land-holder) replies, 'my place. Telangana means my place.'

This visceral connection with Telangana meant that the struggle for secession was perceived as central to the participants' own identity, as well as their survival. The use of the term *desam* highlighted the simultaneous sense of ownership and belonging associated with Telangana, and the centrality of this sense of place to the ontological need for security in the high-risk context of the villages.

This sense of belonging to a territory also served as the basis for asserting ownership of its resources as the means to facilitating the risk negotiation of local people. (This, of course, raises complex questions of justice and difficult normative decisions as to who is legitimately entitled to the world's resources in a context of global risk, an issue which will become more prominent as risk materialises and key resources become increasingly scarce). The visceral struggle for survival and security from risk which the fight for Telangana represented was evident in the passion which the theme of secession aroused and the willingness of villagers to assert their demands for it. Thus, Pavan (male, FC Vaishya, medium land-holder) in Bantala argues, 'we'll fight until we get Telangana.'

The potential for greater representation in a separate state was asserted by Nand (male, BC Yadava, semi-medium land-holder) in Nandanapuram who claims,

> Telangana would be like a smaller family with a better sharing of resources and improved awareness. When the state is small, it can cater to the needs of each and every grouping, look into the detail and serve everyone better.

Nand also asserts, 'the government [of a separate state] could make policies with the consensus of the people and fetch more resources for farmers.' Thus, the demand for secession incorporated concerns with all three dimensions of justice—redistribution, recognition and representation—identified by Fraser (2008: 16–18). However, the demand for recognition and representation was inextricable from the concern with gaining access to resources (redistribution).

Salma (female, SC Madiga, landless) in Nandanapuram asserts the greater autonomy which secession would bring, arguing 'if we separate, we will have our own leaders and be able to do things our own way. We will be doing it for ourselves.' Nirmal (male, BC Yadava, small-holder) in Orgampalle believed that secession would resolve the problem of corruption. He claims, 'now [Coastal] Andhra people are using corruption [to gain access to resources and political influence]. If [the new state of] Telangana comes, it will not be corrupt. We will develop differently.'

In Nandanapuram, Anshul (male, BC Gowda, small-holder) argued that the party system was leading to political expediency and preventing politicians from declaring their true beliefs with regard to secession demands. He claims, '100 per cent *kaavaale*! [We 100 per cent need Telangana]. From a schoolchild in 5th class, to a 60 year old, everyone wants Telangana. They may not be able to say so because of party loyalties, but everyone wants Telangana.'

A small number of villagers, however, expressed cynicism as to whether the new state would permit any change to the negotiation of risk in the region. In Nandanapuram, Nishok (male, SC Madiga, small-holder) claims, 'I hope that employment opportunities will come, but I'm not sure if they will or if it will be the same as it is now.' This related to the awareness of the villagers of the impact of existing power structures on access to resources and democratic practice and the likelihood that local power structures in Telangana would simply replace those of the wider pre-secession state.

Nikhil (male, FC Reddy, semi-medium land-holder) argues,

> everyone wants Telangana. Even I do. But there will not be much change. The *goondas* [criminals] will rise, and there will again be the same thing....I would prefer king's rule to democratic rule.

Ashna (female, SC Madiga, small-holder) in Bantala asserts, 'everybody is saying we should get Telangana, but it won't do anything.' And Ambu (female, SC Madiga, small-holder) in Nandanapuram claims, 'only those who are in politics will benefit from a separate Telangana. No one else.'

In Orgampalle, Pradnesh (male, dominant caste, medium land-holder) argued that the protests were unnecessarily heated and that the voting process itself should be used to bring about a legitimate solution. He claims,

> Right now, there's competition between Telangana and [Coastal] Andhra. Students are staging protests, they're being jailed. There's no need for this. We have the right to vote every five years, so let's make use of that.

Pradnesh's view does not, of course, take account of the way in which the electoral process is itself mediated by power relations, leading to difficulties in challenging the power structure given the influence of that power structure on the political process itself.

Many of the most vulnerable participants were unaware of the relevance of, or were disinterested in, the protests for a separate state. Nipa (female, SC Madiga, landless) in Bantala argues, 'I don't know anything about it [Telangana].' Likewise, Ranjan (male, BC Chakali, landless) in Nandanapuram questions, 'Why would I be interested in Telangana? I don't have anything. I don't understand politics.' And Sajan (male, dominant caste, landless) in Orgampalle questioned my translator: 'people are talking about Telangana. There's much commotion happening. Is it something to do with the elections?'

Democratic Legitimacy and Bt Cotton

This section explores the way in which the legitimation (and delegitimation) of Bt cotton intersects with wider perspectives concerning the legitimacy of the state and the pre-secession state government. This relates directly to the way the legitimation of risk intersects with the legitimation of democracy, as well as the way the power structure of the village intersects with that of the wider state.

Concerns regarding the risks of Bt technology are not channelled through the *gram sabha* and the PRI structure of the democratic process given the diminished epistemic nature of institutionalised democratic practice and the linking of democratic practice to patronage opportunities. Instead, in the case of protestors in Orgampalle and Nandanapuram, their fears for the risks of Bt technology are asserted through their engagement in inter-village mobilisations, an involvement which is organised by the NGOs, Crops Jangaon (Orgampalle) and the Deccan Development Society (Nandanapuram). As highlighted, these NGOs also support and train cultivators in their use of the alternative organic (Orgampalle) and NPM (Nandanapuram) methods and organise for the supply of non-Bt cotton seeds. This involves the co-ordination of the demand for non-Bt cotton seeds across villages.

A number of participants in Nandanapuram and Orgampalle are active in protests and mobilisations against both Bt cotton and Bt brinjal. They are all Backward Caste males from a variety of land-holding categories. In Orgampalle, these include the village elder, Pradnesh (dominant caste, medium land-holder), and Prakash (BC Yadava, marginal land-holder), an NGO employee. In Nandanapuram, both Nand (BC Yadava, semi-medium land-holder), the NPM cultivator and NGO employee, and Rajiv (BC Gowda, small-holder), the Bt cotton cultivator who first introduced the seed varieties to the village, are active in protests against both Bt technology and pesticides.

The absence of NGO involvement for agricultural purposes in Bantala has led to a depoliticisation of the concerns of the villagers regarding the risks of Bt cotton. This has also been influenced by their claims of the improved economic benefits which they assert it has brought (which is belied by their debt exposure), as well as its legitimation by powerful Forward Caste villagers, such as Pavan.

The depoliticisation of the risks associated with Bt technology in Bantala should not be taken to mean that the village was depoliticised in other areas. The protests in which villagers engaged, however, were largely directed by Pallav, the *sarpanch*'s husband. The village servant, Neelam, for instance, was instructed by Pallav to attend a protest in a nearby town concerning the shortage of fertilisers during the research period.

Similarly, during one field trip, I was intrigued to find a local bus stranded in the centre of the village surrounded by villagers, including Pallav. The group refused to allow the bus driver to leave until they had secured his manager's agreement (by mobile phone) that the village would be provided with a second daily bus service. As has been noted, the village was also actively engaged in hunger strikes for a separate state, again organised by Pallav. It is to be assumed that the engagement of villagers in protests against Bt cotton would have challenged the authority of both Pavan (male, FC Vaishya, medium land-holder who introduced the technology to the village) and Pallav (given Pavan's power in the village and his support for Bt technology).

There were, nonetheless, indications that the legitimation of Bt cotton in Bantala was waning, a change which was being negotiated through the traditional village power structure. This relates to the report by Crops

Jangaon that village elders from Bantala had expressed an interest in visiting Orgampalle to ascertain whether organic methods might be a more legitimate cultivation method than Bt cotton to negotiate agrarian risk. The account illustrates the ongoing role of elders in representing the village externally. It also highlights the role of the NGO as an intermediary between the villages. The current research suggests, however, that any potential change to cultivation method which the elders may suggest would need to be legitimated by Pallav and the powerful Forward Caste cultivators, most notably Pavan, before it could gain more widespread acceptance in Bantala.

Protests for a separate state and those opposing Bt technology operated outside of the institutionalised structures of democracy in ways which sought to challenge power relations. Unlike the protests for a separate state, however, which focussed on inter-regional inequality and transcended local power differentials, protests against Bt cotton seek to gain recognition for the inequality within Telangana itself. These protests, it is argued, will continue to focus attention on democratic legitimacy within the new state as part of the concern with power relations and the differentiated access to resources, exposure to risk and limited ability to gain recognition and representation within the political process which this concern entails.

Figure 7.1 explores the relationship between perceptions of state legitimacy, in terms of the perceived recognition by the state of the risk exposure of cultivators and the legitimation of Bt cotton. This is indicated in the differing answers provided by villagers in response to the question: 'The government is interested in the experience of farmers who grow Bt cotton.' (See Appendix 17, Q. 11).[6]

As can be seen from Fig. 7.1, the highest degree of satisfaction with the government's recognition of Bt cotton cultivators is in Bantala, where half of the participants strongly agree with the statement that the government is interested in their experience. Chitta (male, dominant caste, semi-medium land-holder) strongly agrees given his perception of the government's support for the technology. He argues, 'the government is praising those who opt for [Bt] cotton. Cotton farmers are believed to be richer farmers.' The claim highlights that, for Chitta, the government's legitimacy is enhanced through its promotion of a technology which, he believes, increases cultivator yields and incomes.

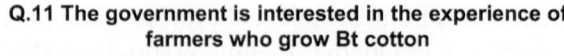

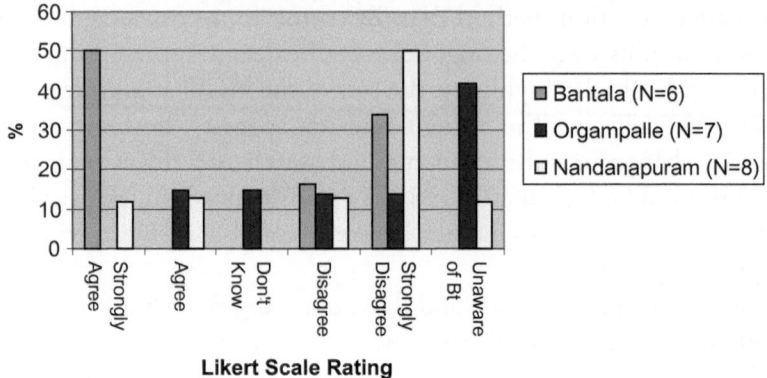

Fig. 7.1 State recognition and Bt cotton cultivation

The sense of recognition for their plight is also asserted by many participants in Bantala given their satisfaction with the purported economic benefits which they argue Bt cotton has secured for them and which the pre-secession state was seen as supporting. Hence, the legitimation of Bt cotton coincided with the legitimation of the economic policies of the pre-secession state government in Bantala.

The changing bases for the legitimation of Bt cotton in Bantala, however, and the implications of this for state legitimacy are indicated by the fact that the other half of participants in Bantala disagree or strongly disagree that the government is interested in their experience. This is given their concerns for the declining yields, increasing costs and fears for the future associated with their cultivation of Bt cotton. Ashna (female, SC Madiga, small-holder), for instance, argues, 'they [politicians at election time] come, but they are not really concerned. It really relies on the *sarpanch*, not the government.'

The delegitimation of Bt cotton in Nandanapuram is associated with a strong sense of misrecognition by the state in relation to its efforts to alleviate their risk exposure. In Nandanapuram, the government is delegitimated on the basis of its failure to understand the real needs of cultivators. This leads half of the participants to strongly disagree that the government is interested in their plight. Rajiv (male, BC Gowda, small-holder) claims, 'for politicians, Bt cotton is a small problem, so they don't have time to listen.' Ambu (female, SC Madiga, small-holder) also argues, 'when we lose our crop, the government is not interested.'

Nand (male, BC Yadava, small-holder) strongly disagrees that the government is interested in Bt cotton cultivators. He asserts, 'it's the farmer who is realising the effects of Bt [cotton] and then trying to protest but, because of America, and the companies who produce the technology, and the World Trade Organisation, the government [here, the central government in Delhi] is not able to say no to this technology. India is kind of trapped.'

Nand (male, BC Yadava, small-holder) argues that the wider global power structure in which the Indian government is embedded is having a direct impact on state legitimacy, claiming 'it's a flaw in society that they [the Indian government] don't respond to the petitions they get....Instead, they are using *lathi*[7] charge to repress the protests.' He also notes the epistemic injustice associated with democratic practice, arguing 'you can only prove that Bt cotton is toxic if you can scientifically prove it through research.'

Among those who are aware of what Bt cotton actually is in Orgampalle, perspectives are more varied. This is given the fact that, although cultivators in Orgampalle delegitimate Bt cotton, they also feel recognised given that they perceive their participation in protests to have contributed to the moratorium on Bt brinjal. Prakash (male, BC Yadava, marginal land-holder) notes the efforts of the former Minister for the Environment and Forests, Jairam Ramesh, to engage with cultivators, arguing 'Jairam Ramesh has conducted a big meeting for all the farmers some time back.'

Nonetheless, Prakash still argues that not enough is being done to address the risk exposure of cultivators given his belief that Bt technology should be banned. He claims, 'there are thousands of farmers fighting on the street against Bt [technology]. Why does the [Indian] government not listen to them?' Like Nand, Prakash also notes the epistemic injustice to which cultivators are subject given that they are unable to provide scientific evidence for their assertions of the risk of Bt cotton. He argues, 'our limitation is we can't analyse scientifically.'

Figure 7.2 explores the perspectives of villagers with regard to the success of the democratic process in bringing about the 'passive revolution'

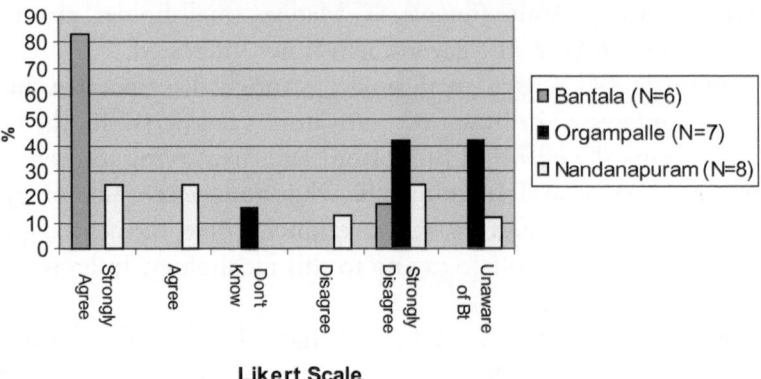

Q. 14 The democratic process has led to greater equality of opportunities in Indian society

Fig. 7.2   Views on the legitimacy of the democratic process

(Corbridge and Harriss 2000: 21) envisaged by Nehru in terms of challenging the power structures of Indian society and promoting greater equality.

The legitimation of Bt cotton in Bantala appears to coincide with a high degree of legitimacy for the ideal of democracy. Despite the fact that the village is characterised by the presence of a strongly dominant caste, inequality of access to resources and strict segregation, the majority of participants strongly agreed that the democratic process is leading to greater equality. This general perception of improvement among all castes contributes to the ongoing legitimation of the (highly oppressive) village power structure.

The perception of greater equality was particularly asserted by the powerful Forward Castes, Pavan (male, FC Vaishya, medium land-holder) and Sudhakar (male, FC Reddy, semi-medium cultivator), as well as the dominant caste participant (Chitta, BC Kuruma, semi-medium cultivator). For these participants, democratic practice is perceived as contributing to a more equal society even as they are more powerful than others within it and benefit disproportionately from it.

Ashna (female, SC Madiga, small-holder) was the only participant in Bantala who strongly disagreed with the statement that the democratic process was leading to greater equality. She argues, 'opportunities are not equal. Scheduled Castes and Backward Castes are being constrained. Even if someone from those castes knows better, [upper castes] would oppress him [*sic*].'

In Nandanapuram, where power is contested and Bt cotton is delegitimated, the perspectives of participants on the contribution of the democratic process to challenging power relations and securing greater equality are more diffuse. Rajiv (male, BC Gowda, small-holder) strongly disagrees that there is greater equality, arguing 'in this democracy, there are *goondas* [criminals] disguised as politicians.' He also asserts that the democratic process cannot function properly unless people are educated. He claims,

> Only if everyone is educated, will they know the power of the vote. Now the politicians give money to the illiterate to get votes. Only when everyone is educated, will people know what the vote means.

Nikhil (male, FC Reddy, semi-medium land-holder) in Nandanapuram strongly agrees that the democratic process has led to a greater emphasis on equality. He does not, however, regard this as necessarily positive given the absence of a final decision-making authority and his preference for hierarchy. He claims, 'I would prefer king's rule to democratic rule. A least a king would ban things [like Bt cotton].'

He also highlights the difficulty in securing representation and recognition through the Indian democratic process, claiming

> when politicians are looking for your vote, they treat you like gods. Then when you go to seek assistance in your parliament, he [the politician] doesn't even remember which village you came from.

In Orgampalle, Pradnesh (male, dominant caste, medium land-holder) claims,

> I don't think there are equal opportunities at all. If a rich person goes to a politician, or is competing for something, he gets it. Someone like me who is poor would never get an opportunity in front of him.

Prakash (male, BC Yadava, marginal cultivator) also strongly disagrees that there is greater equality as a result of the democratic process. He asserts the failure of the Indian state to respond to the protests of Bt cotton farmers due to the perceived favouring of the companies who promote the technology. He claims,

Democracy is not one man's say or one company's say, but the say of everybody..... [The Indian government] only favours the [multinational] companies [who market Bt cotton].

Pradnesh (male, dominant caste, medium land-holder) also argues that cultivators are vulnerable given their weaker position in relation to wealthy multinational companies. He claims, '[farmers] are being cheated by the seed companies. At any time, the farmer is in debt always.' He also notes, however, the influence of the media in contributing to greater equality in democratic practice. He argues, 'what is changing is the development of the media. Today, we get more information and people are more knowledgeable. This is helpful for fighting for equality.'

The findings suggest that the legitimation of Bt cotton within the concentrated power structure of Bantala coincides with the legitimation of democratic practice with regard to the perception that it promotes greater equality. This perspective is linked to the perception of the general increase in wealth in the village, a view which coincides with pervasive debt levels, particularly among the most powerful.

The perception that the democratic process has contributed to greater equality in Bantala is asserted most strongly by participants from the Forward Castes and dominant caste who are power-holders in the village and, in the case of Chitta and Pavan, have previously asserted a belief in the legitimacy of the inequality associated with caste (see Chitta's quote in Chap. 6: 'as fingers on the same hand are all unequal, so castes are not equal.'). This suggests that life in Bantala is driven by the interests of the powerful and supported by government initiatives (which contribute to welfare enhancement) in ways which mask, rather than challenge, the underlying power dynamic and the indebtedness with which this is associated.

Opposition to Bt cotton, on the other hand, is associated with greater critical awareness of the inequality of power relations and the way this mediates democratic practice. In Orgampalle and Nandanapuram, this awareness of power relations gives rise to a perceived inability of institutionalised democratic practice to challenge the inequality which serves as its foundation in the absence of extra-institutional protests. These views relate to democratic practice at both state and national levels.[8]

The protests against Bt cotton create a somewhat broadened sense of solidarity given their inter-village nature, as well as the involvement of different categories of land-holder. Although not apparent from the participants in the current study, the protests also increasingly include females and Scheduled Castes.[9] As such, mobilisations against Bt technology represent the potential for an emergent solidarity with regard to their risk exposure among Backward and Scheduled Caste males and females with land holdings of below 20 acres. The extent to which females, Scheduled Castes and the landless, as vulnerable groups, engage in challenging their risk exposure as part of wider political mobilisations in the rural areas appears dependent upon the nature of the NGO intervention in particular villages. As highlighted, engagement with NGOs is a decision which must itself be legitimated through the village power structure and, therefore, relies upon the assessments of power-holders in the villages. The political involvement of these vulnerable groups is also, however, determined by the degree to which such engagement is legitimated as appropriate and feasible by these groups themselves, following assessment of the likely impact upon their already heightened exposure to risk.

This chapter has explored the way in which the institutionalised practice of village democracy is strongly influenced by the village power structure and linked to the ability to secure patronage. This limits its deliberative, epistemic dimension and serves to reinforce the power structure and suppress dissent, given that power-holders also control the dispensing of patronage. The chapter has examined the way in which demands for secession were strongly related to attempts to assert discrimination in the inter-regional allocation of resources and the failure of the pre-secession government to adequately recognise and represent the heightened risk exposure of the Telangana region. Given their focus, these demands paid less attention to the power differentials within Telangana itself.

The analysis also highlights the strong association between perspectives on Bt technology and state legitimacy, with opponents of the technology being more likely to assert a sense of lack of recognition by the state, and to delegitimate institutionalised democratic practice. The division in Bantala regarding the sense of recognition by the government (Fig. 7.1) suggests a growing dissatisfaction among villagers in relation to the government's attempts to address their exposure to agrarian risk. This coincides with a waning legitimation for Bt technology as the approach supported by the government.

While in Orgampalle the perceived lack of recognition by both the state and central governments was somewhat alleviated by the series of public consultations held by Jairam Ramesh with regard to Bt brinjal, participants in the village also asserted their delegitimation of democratic practice due to the impact of power relations on its operation. The wider power relations associated with the Indian state were asserted by participants as leading the Indian government to favour the interests of multinational seed companies over those of cultivators.

Notes

1. Interviews were conducted with the *sarpanches* in Orgampalle (13/2/2011) and Nandanapuram (22/11/2010). The latter, a female, had her husband present throughout. Despite a number of requests, the female *sarpanch* in Bantala refused to be interviewed.
2. Akhil (male, dominant caste, marginal cotton-holder) notes that the *sarpanch* for the village was previously female and claims, 'there is space for good people to become *sarpanch*.'
3. These stipulations vary by state. The Andhra Pradesh requirements pre-secession are available at: http://www.rd.ap.gov.in/EGS/SA_Rules_170408_final.pdf. Section 3 (2): q. Accessed on 30/3/2017.
4. Unfortunately, the opportunity to attend the meeting in Nandanapuram and Orgampalle did not present itself.
5. The world's largest masonry dam, the Nagarjuna Sagar, is located in Telangana.
6. A reviewer notes that the phrasing of this question is problematic. This relates to the fact that what is meant by the government being 'interested' is unclear, as is the level of government which respondents were envisaging in their responses. The critique is a valid one. It is mitigated to some extent, however, by the fact that the questionnaire was completed as part

of my interviews with participants. The questions, therefore, served as a starting point for discussion, and respondents generally provided clarification on the reasons for their answers. Only selected excerpts of the responses are provided here. In terms of the level of government, this was often clarified as the state government; on rare occasions, however, references to global players or to the state of India highlight that it is the wider macro of the central Indian government in Delhi to which the respondent is referring (as in Nand's response, for instance). Being 'interested' was taken to mean that politicians listened to and understood the concerns of cultivators, a perception which was reinforced through their visits to the villages. The latter was something which many participants noted occurred only when candidates were looking for votes.

7. A *lathi* is a long stick used by the police to dispense large crowds.

8. As highlighted in Chap. 3, the protests against Bt cotton are associated with different levels of engagement, depending on where their demands for recognition are focussed. Generally, inter-village protests which are located within a particular state demand recognition from the state government. Protests which involve participants across a number of states often seek wider recognition from the central government in Delhi, as well as the federal state itself.

9. A mobilisation against Bt cotton in 2012 involving 5000 Scheduled Caste females from 75 villages in pre-secession Andhra Pradesh was co-ordinated by the Deccan Development Society. Available at: http://www.gmwatch.org/index.php/news/archive/2012/13848-women-march-for-total-ban-on-bt-cotton. Accessed on 30/3/2017.

Bibliography

Corbridge, S., & Harriss, J. (2000). *Reinventing India: Liberalization, Hindu nationalism and popular democracy*. Cambridge: Polity Press.

Fraser, N. (2008). *Scales of justice: Reimagining political space in a globalizing world*. Cambridge: Polity Press.

Henrich, J. (2001). Cultural transmission and the diffusion of innovations: Adoption dynamics indicate that biased cultural transmission is the predominate force in behavioural change. *American Anthropologist, 103*(4), 992–1013.

Michelutti, L. (2007). The vernacularization of democracy: Political participation and popular politics in North India. *Journal of the Royal Anthropological Institute, 13*, 639–656.

Powis, B. (2003). Grass roots politics and 'second wave of decentralisation' in Andhra Pradesh. *Economic and Political Weekly, 38*(26), 2617–2622.

Robinson, M. S. (1988). *Local politics: The law of the fishes: Development through political change in Medak district, Andhra Pradesh (South India)*. Delhi: Oxford University Press.

Srikrishna, B. N., Duggal, V. K., Singh, R., Shariff, A., & Kaur, R. (2010). *Committee for consultations on the situation in Andhra Pradesh*. New Delhi: Government of India.

8

Legitimation in a World at Risk: Lessons from the Villages

In this chapter, the findings of the village research are discussed. The chapter explores how both the legitimation and delegitimation of Bt cotton are mediated through village power relations. In the case of Bt cotton cultivators, the research finds that it is the more powerful participants who introduced the technology to the villages who are among the most indebted of the study. In Bantala, this indebtedness relates to the way risk-taking is legitimated by Forward and dominant castes; in Nandanapuram, the indebtedness serves as a constraint which locks participants in to Bt cotton's ongoing adoption, despite its delegitimation.

The chapter also examines the legitimation of democracy, exploring how both the *gram sabha* and the electoral process are associated with patronage opportunities. These are mediated through village power relations in ways which serve to reinforce the power structure. The protests against Bt cotton which are coordinated by NGOs operate outside of institutionalised democratic practice. These not only challenge Bt technology as a response to agrarian risk; they also seek to extend the right to justification in ways which attempt to enhance democratic legitimacy. This book argues that the issue of GM crops, which was so

© The Author(s) 2018
E.L. Desmond, *Legitimation in a World at Risk*,
https://doi.org/10.1007/978-981-10-6065-6_8

central to the pre-secession state's attempts to secure its precarious legitimacy, will now become pivotal to attempts by the new state government to establish its legitimacy.

The Legitimation of Risk and Village Power Relations

The analysis indicates that, in the high-risk context of Warangal, the ambiguous risk of Bt cotton is traded against its potential economic benefits as part of the process of its legitimation. As the research highlights, however, all three villages have assessed this trade-off differently. This book argues that this variation results from differences in the village composition and resulting power arrangements, as well as the views of power-holders. These aspects influence the way in which the 'social risk positions' (Beck 1992: 40) as defined here by land-holding, caste and gender are differentially impacted not only in their exposure to risk (given differences in access to resources) but also in their ability to contribute to risk construction (arising from differences in the ability to gain recognition and representation for risk exposure).

The concentration of power in Bantala is evident from the way access to key resources is restricted to dominant and Forward Castes (see Appendix 11). The suppression of the Scheduled Castes in the village, many of whom are wage labourers, is evident from labour rates which were below the official minimum. These lower rates for agricultural labour favoured cultivators, but exacerbated the risk of the landless, who relied on daily wage labour for their subsistence. The legitimation of Bt cotton on the basis of its contribution to wealth creation in the village, even while most of the participants were significantly in debt, suggests that indebtedness is itself legitimated as part of the potential for wealth creation which Bt cotton represents.

The legitimation of the approach to agrarian risk associated with Bt cotton is reinforced in Bantala through the conspicuous consumption which indebtedness funds, as well as the perception of increased wealth which such consumption suggests. The particular tendency to risk-taking among the aspirational Forward Castes in the village is evident, not only from their high debt levels (Table 6.4) but also from their high-risk strategy of leasing land as the means to securing higher incomes (Appendix 11). The level of indebtedness of Forward Castes in Bantala is almost matched by that of small-holders in the village.

The risk which Bt cotton entails in terms of its higher costs is adopted by small and marginal cultivators in Bantala not only because of the tendency towards Sanskritisation (Srinivas 1966: 6) but also because the technology has, at times, been associated with higher yields and incomes. The analysis highlights that these benefits are asserted by Forward Caste and Scheduled Caste participants alike. This view coincides, however, with pervasive indebtedness (only Sudeep: male, SC Mala, marginal cotton-holder, was debt-free). As highlighted, both of the marginal cultivators in Bantala have access to non-farm incomes which serves to alleviate their exposure to indebtedness (see Appendix 11). However, the conspicuous consumption of the powerful serves to reinforce the view of greater prosperity generally, even while this is being funded by borrowing.

The study also highlights that, in Bantala, the legitimation of the inequality associated with the power structure by the less powerful Scheduled Caste participants is precarious and based upon their perception that they are better off than previously. It is worth noting, however, that the perceived improvements which the Scheduled Castes in Bantala highlight cannot be attributed to Bt cotton alone. They are also the result of government initiatives such as minimum labour wage rates, the Indiramma housing scheme, free electricity for cultivators, NREGS and the Public Distribution System. In the case of Bantala, these programmes, and the perception of improved risk negotiation with which they are associated, serve to legitimate the village power structure and, in so doing, contribute to a depoliticisation, not only of the inequality of access to resources in Bantala but also of the perceived risks of Bt cotton given that it is the approach favoured by the apparently wealthy village power-holders.

The large size of Nandanapuram and the absence of an overriding authority or dominant caste contribute to a contested power structure. This is characterised by an underlying sense of tension, as well as indecision in relation to cultivation practice. Here, the certainty expressed by

Bt cotton cultivators that the technology is linked to animal deaths and indebtedness is traded against their desperate need for higher crop yields to alleviate this indebtedness.

While Rajiv, who introduced Bt cotton to the village, is not himself a significant land-holder (he owns two acres and leases a further two), he is a large, gregarious male whose caste (BC Gowda) is relatively powerful in the village (see Appendix 5). His position of power is evident from his ownership of a borewell and his tendency to risk-taking as the means to prosperity given his leasing of land (Appendix 12).

The conflicted position of Rajiv, who continues to cultivate Bt cotton even as he protests against it, results from his significant exposure to debt (Rs 250,000). This arose from dowry payments, as well as cultivation costs. He is now constrained from changing cultivation practice due to this indebtedness and his hopes for a bumper yield which will allow him to alleviate some of his debt. His position illustrates the view of Kumbamu (2007: 891) that '[m]any desperate Warangal farmers continue cultivating Bt cotton (so-called white gold) in the hope that one good crop may help them out of the debt trap and the death trap.'

Rajiv's ongoing adoption of Bt cotton occurs despite the fact that an alternative NPM practice exists which could easily be adopted given the extension network which is already available for this within the village.[1] Environmental concerns are traded, however, against Rajiv's need to negotiate the added dimension of risk associated with his indebtedness and the gamble to secure the potentially higher yields which Bt cotton represents.

The relatively weak caste position of the NPM cultivator, Nand (Nand's caste, the BC Yadavas, own just over 2 per cent of village land: see Appendix 5), as well as the debt levels of villagers, has limited the appeal of NPM practices. Nand is, nevertheless, a focal point for NGO engagement within the village. Not only is he university educated and an NGO employee (SEED), an occupation which allows him to secure non-Bt cotton seeds; he also works closely with the Deccan Development Society who provide extension services for NPM practices in the village.

Nand's NGO employment and affiliation, and the increased wealth which his employment and farming practice support, serve to bolster his position in the village. This has led to a small percentage of villagers (10 per cent) emulating the practice which he advocates. It has also meant that Rajiv (male, BC Gowda, cotton small-holder), who initially introduced Bt cotton to the village, has joined with Nand in protesting against Bt technology and pesticides, even as he continues to adopt both because of his indebtedness. While this strategy represents his hopes for high yields, it is undertaken with ongoing environmental concerns associated with his cultivation practice. It also exposes Rajiv to the risk of further indebtedness in the event of crop loss given the high costs involved.

As Table 6.4 highlights, it is again the medium land-holders and Forward Castes who are the most indebted in Nandanapuram. As in Bantala, this again suggests the centrality of these categories in spearheading the 'revolution of rising expectations' (Gupta 2002: 86) associated with the Vision 2020 neoliberal development model in which Bt technology was embedded in the pre-secession state.

The small size of Orgampalle, and its caste homogeneity, has allowed Pradnesh, the charismatic village elder, in conjunction with Crops Jangaon, to promote a more collective, cohesive approach to risk negotiation. The legitimation of this approach is facilitated by the absence of an aspirational, highly indebted, Forward Caste population in the village. The influence of the elder in Orgampalle is clear from the fact that many of the participants are unaware of why Crops Jangaon and Pradnesh object to Bt cotton; they are simply refusing to adopt it given their legitimation of Pradnesh's authority.

Because villagers in Orgampalle believe that the village as a collective is more economically secure as a result, they continue to legitimate Pradnesh's power. This means that the enhanced access to resources which Pradnesh enjoys relative to the other villagers (see Appendix 13) remains unchallenged. The practice of sharing assets, such as oxen, and keeping cultivation costs, including labour, to a minimum, means that the average indebtedness in Orgampalle is the lowest in the study. Similarly, Pradnesh is debt-free and opposes consumerism and the individualisation of risk negotiation. His influence in this regard has not, however, eliminated borrowing entirely in the village.

Indebtedness in Orgampalle is instead associated with aspects other than cultivation, such as house construction, dowry, labour charges and health (see Appendix 16). The situation in Orgampalle highlights that Bt cotton is not the only factor in rural indebtedness. However, the differences in debt levels between organic cultivators in Orgampalle and Bt cotton cultivators in the other villages suggest that Bt cotton cultivation is associated with an added dimension of financial risk. On the one hand, this is associated with the conspicuous consumption of Forward Castes, but it is also linked to the additional costs of cultivation associated with Bt cotton itself. As highlighted, the wider study on which this book is based showed that Bt cotton cultivators spent, on average, Rs 6740 per acre on the seeds, pesticides and fertilisers specifically related to Bt cotton cultivation in the 2010/2011 season.

The analysis highlights that, while the marginal cultivators in Bantala have been able to alleviate their risk exposure given their access to off-farm employment, debt exposure is generally significant for Bt cotton cultivators with small-holdings in both villages and marginal holdings in Nandanapuram. This is particularly high risk for these categories given their limited access to assets to sell in order to clear debts in the event of crop loss. This risk is exacerbated by the fact that their restricted access to land often coincides with a lower caste and/or female status associated with a weaker bargaining position within the power structure. As highlighted in Chap. 4, small and marginal land-holders continue to be most at risk of suicide in the new state, accounting for 55 per cent of farmer suicides in Telangana in 2014 (Goyal 2015: 283).

As Chap. 4 also noted, however, the extent of the risks which semi-medium and medium cultivators are taking is indicated by the increasing farmer suicide numbers among these categories of land-holder in Telangana (Goyal 2015: 283). The rising incidence of suicide among these larger land-holders suggests the increasing non-viability, not only of cultivation but also of the legitimation of indebtedness as the means to progression which this study suggests these categories are often spear-heading.

Table 6.4 indicates that the exposure to debt among females in Orgampalle and Nandanapuram is greater than for males. As has been highlighted, Telangana has the highest number of female farmer suicides in India (Goyal 2015: x). Given the lack of access to land title, females are also often obliged to borrow from non-institutional sources. In the current study, two female participants, Ashna (female, SC Madiga, marginal cotton cultivator) in Bantala and Aruni (female, dominant caste, marginal land-holder) in Orgampalle, have borrowed from Micro-Finance Institutions. This source of borrowing heightens their exposure to risk given the high interest rates and coercive approach to debt recovery associated with these institutions.

The widespread diffusion of Bt cotton is supported by seed dealers who no longer stock non-Bt seed varieties. While the seed dealer I interviewed argued that the non-availability of Bt cotton was due to lack of demand for non-Bt seed varieties, cultivators highlighted that seed dealers are influential in promoting the technology, not least due to the higher commission which it earns for them. Given the difficulties with obtaining non-Bt cotton seeds, alternative practices entail the need for NGO engagement. This potentially locks cultivators into a power relation beyond the village due to the fact that many NGOs come with their own philosophies and assert their own particular views on cultivation.

The potentially changing basis for the legitimation of Bt cotton in Bantala is indicated by the report from Crops Jangaon that village elders from Bantala have expressed an interest in visiting Orgampalle to assess the legitimacy of organic cultivation. It is clear, however, that any change to the cultivation method in Bantala will need to be legitimated by Pavan and the other Forward Caste and dominant caste land-holders before it will gain more widespread acceptance. This indicates how the traditional power structure associated with village elders co-exists with the power relations as determined by land-holding, caste and gender and, in some ways, serves as a check to their legitimacy in the negotiation of risk.[2]

The study highlights the way in which the legitimation of risk involves an 'assessment of trade-offs' (Renn 2008: 196). This entails an evaluation of 'how much uncertainty one is willing to accept for some future opportunity' (ibid.). It also illustrates, however, the way this assessment is mediated through power relations in local contexts—here, an Indian village.

Beetham (2013: 43) claims that power relates to 'the ability to influence or control the actions of others.' The use of prestige and conformist biases in the analysis explores the impact of power relations on the choice of cultivation method. This highlights that the social learning associated with these biases does not entirely preclude environmental learning; Bt cotton cultivators in both Bantala and Nandanapuram first observed the performance (in terms of yield) of Bt cotton among the early adopters (Pavan in Bantala and Rajiv in Nandanapuram) before taking the decision to adopt it themselves; participants who cultivate Bt cotton in both villages also assert their concerns for the environmental risks (such as damage to soils and animal deaths) to which they believe the technology is somehow contributing.

The fact that cultivators in both Bantala and Nandanapuram continue to adopt Bt cotton, despite these concerns, illustrates the 'agricultural deskilling' identified by Stone (2007: 72). The prevalence of Bt cotton, and the regime of pesticides and chemical fertilisers with which its cultivation is associated, has limited the ability of cultivators to respond individually to the highly complex environment associated with dryland agriculture. However, there are subtle differences between Bantala and Nandanapuram in this regard.

Bt cotton cultivators in Bantala are uncertain of Bt cotton's links to environmental risk and argue that the technology may or may not be associated with damage to the soils - although greater certainty is expressed with regard to its contribution to animal deaths; in Nandanapuram, however, cultivators express greater certainty of the risks of Bt cotton than participants in Bantala—but they also assert their inability to change given their debt levels and the widespread adoption of the crop by everyone else. For these reasons, cultivators in both villages fail to act upon their suspicions (Bantala) and perceived certainty (Nandanapuram) of risk. Hence, their learning in relation to risk is mediated through social and economic factors in ways which lead to a reluctance, in both cases, to alter behaviour in line with concerns. This hesitation is associated with a perception of their limitations with regard to cultivation and the lack of availability of options which are perceived as having the same yield potential. (This is despite the finding that in the far from unusual catastrophic 2010/2011 season it was an NPM cultivator - Nand - who received the highest yield).

The importance of the power structure in such situations of uncertainty is highlighted by Renn. Renn (2008: 28) notes that, in contexts marked by the disorientation and ambiguity associated with risk, the power structure itself provides significant ontological security. The centrality of the power structure to the way in which risk is constructed and negotiated in the villages is illustrated in the analysis. The role of the powerful with regard to risk negotiation is also noted by Cooke (2001: 107) who argues that 'risk-taking is a cultural value' with 'risky individual[s] being the most influential' (ibid.). The research indicates the way the social relations of the particular context have a significant impact on how risk is constructed and negotiated as part of its legitimation.

This social aspect to the construction of risk is evident in the differences in perceptions between Bt brinjal and Bt cotton in Bantala. The perceived trade-off associated with Bt brinjal is markedly different to that for Bt cotton for all participants in the village given that Bt brinjal is a food crop. This is despite the fact that cotton oil is widely used in cooking.

Beetham (2013: 43) argues that 'the power a person has indicates their ability to produce intended effects upon the world around them.' The current research provides an insight into how the legitimation of risk involves a legitimation of the power structure in ways which contribute to agricultural deskilling and impact upon perceptions and priorities in the assessment of trade-offs. The legitimation of risk in Orgampalle, and its mediation through a power structure which is defined by the values of a charismatic elder, has led to the same environmental concerns which were evident in Bantala and Nandanapuram being negotiated differently.

The situation in Orgampalle highlights Kumbamu's (2007: 891) critique of Stone that 'experimentation is occurring' in Warangal leading to the rejection of Bt cotton, even while Bt cotton cultivation is pervasive. The adoption of organic cultivation in Orgampalle is supported by Pradnesh's decision to align with Crops Jangaon, an association which links the village to a network beyond the village in order to alleviate any potential risks of this alternative and marginal practice. This strategy is particularly astute given that Orgampalle is being used as a model to illustrate the benefits of organic practice, a fact which means that both Crops Jangaon and Pradnesh have a vested interest in the village's success in this regard.

Beetham (2013: 59) argues that a given power structure is 'justified when it can be shown to serve not merely the interests of the powerful, but those of the subordinate also.' Through the legitimation of a successful approach to risk which is promoted by power-holders, the power structure is itself legitimated. This legitimation is likely to continue as long as there is no disruption to the perception by the less powerful of the relative material benefits they are experiencing as a result of the strategy which is supported by the power structure.

Habermas ([1973], 1976: 96) argues that '[l]oyalty may be hypocritically simulated…or carried out in practice for reasons of material self interest.'

This was evident in Bantala in the case of the two Scheduled Caste marginal cotton-holders (the female, Ashna, and male, Sudeep). Both of these participants expressed their awareness of the suppression of the lower castes in the village but tolerated it given their perception that they were relatively better off than they had been previously. This perception served to avert the type of legitimacy crisis described by Beetham (2013: 168) which would occur 'when there [was] a serious threat or challenge to the rules of power, or a substantial erosion in the beliefs which provide their justification.'

Given the erratic crop performance associated with agriculture in India, the stability of the perception of the relative economic well-being among the vulnerable is far from guaranteed. As Rao and Suri (2006: 1552) argue, '[i]t is not the crop loss in one year but recurrent losses that ruin the economic conditions of farmers.'[3] The precarity of the economic situation means that local power structures are also under constant threat of challenges to their legitimacy. Thus, the risks associated with recurrent crop loss, particularly with a high-cost method such as Bt cotton, carry not just an economic but also a significant social cost given the contribution to unrest which they may entail in such a context.

The potential visit of village elders in Bantala to Orgampalle highlights the view of Abromeit and Stoiber (2007: 37) that 'both the needs and beliefs within societies…may change.' This means that power-holders must be ever vigilant to risk negotiation as a concern for the legitimacy of their own positions in the villages. In this way, the power-holders who take the lead in negotiating risk are themselves reliant on their responsiveness, over time, to environmental cues in risk negotiation as the means to safeguarding the legitimacy of their own positions.

This aspect of power, and the need for power structures to legitimate themselves in local contexts, somewhat mitigates the potential for agricultural deskilling to sustain itself longer-term. This suggests that social and environmental learning have the potential to draw closer together over time as risks become apparent as material realities and power-holders fear for their positions. Of course, in a context of pervasive and acute exposure to risk, there is also the potential for a more authoritarian approach to emerge which asserts the social dimension of power relations over and above environmental cues.

Within this, the power structure of the new state government of Telangana would be well advised to pay heed to the findings of the

ICRISAT study of Walker and Ryan (1990: 355). This asserts that crop improvement technology in dryland regions must be regionally specific given the significant inter-regional differences in climate and soil. This natural variability means that the blanket adoption of a high-cost innovation, such as Bt technology, in a dryland context, like Telangana, will always have winners and losers. Given the often tenuous legitimacy of village power structures which this study has highlighted, it is clear that an approach to agrarian risk which delivers benefits disproportionately and erratically in an unequal and precarious context not only exacerbates the risk of cultivators but also of political instability itself.

The Legitimation of Democracy and Village Power Relations

With regard to the legitimation of democracy, the analysis illustrates how the historical struggle to secure a legitimate exercise of power in Telangana continues to influence demands for the right to justification in the region. This most notably concerns the differentiated access to the key resources required to negotiate agrarian risk—namely, water, land and education as the means to securing off-farm employment.

As illustrated, the majority of participants asserted their perception of injustice in relation to the inter-regional distribution of resources, and the failure to adequately recognise and represent the particular risk exposure of Telangana, as the basis for their secession demands. The analysis also highlights, however, that a number of the most vulnerable landless participants were unaware of what the struggle was for. A small minority also argued that a new state would not alter the distribution of resources given the deeply embedded power structure within the region itself.

The examination of the *gram sabha* and electoral process, particularly in relation to the key position of village *sarpanch*, illustrates the impact of power relations on the way democratic practice is legitimated in the villages. The analysis highlights how the linking of democratic practice with patronage opportunities often serves to compromise its deliberative aspect. It is only in Orgampalle that a participant highlights this problem-solving component of the meeting (Akhil: male, dominant caste, small-holder).

The research also indicates, however, that the distribution of patronage is mediated through village power relations in ways which link the legitimation of democracy directly with the legitimation of risk. The way democratic practice is legitimated in the villages in this regard is illustrated through an exploration of perspectives on the *gram sabha*. The meeting has come to represent the central forum for the accessing of resources to negotiate risk; given, however, that it is mediated through village power relations, it does not problematise the differentiated exposure to risk of the villagers, nor challenge the impact of the power structure on the way in which patronage is differentially distributed to address such risk.

The analysis highlights that the *gram sabhas* in both Bantala and Nandanapuram are poorly legitimated. My own attendance at the meeting in Bantala confirmed that it is also highly gendered. The presence of a dominant caste in Bantala meant that the meeting was dominated by the *sarpanch*'s husband, Pallav. While an opportunity to attend the meeting in Nandanapuram and Orgampalle did not present itself, interviews with most female participants in these villages confirmed that they viewed their attendance at the meeting as inappropriate (not their 'level') or impossible due to their household duties.

Given the gendered nature of the *gram sabha*, Ambu (female, SC Madiga, small-holder) in Nandanapuram asserted her prioritisation of Self-Help Group (SHG) meetings. Although loans from SHGs are low interest and the lending is founded on a co-operative approach to risk, the situation with Salma, the landless, SC Madiga participant in Nandanapuram, highlights the need for caution in relation to borrowing as the means to alleviating the risk of the vulnerable. Salma, a widow with two children who lives with her parents, is Rs 10,000 in debt to an SHG due to the failure of her paddy crop on leased land (Appendix 15). She must now pay off this debt from her agricultural wages and meagre widow's pension, in addition to meeting her other financial commitments.

The *gram sabha* is regarded with disdain by upper castes who assert their caste status as the basis for their delegitimation of politics (although they were notably vocal in asserting their need for a separate state). Marginal and landless participants also did not attend the meeting. Sudeep (male, SC Mala, marginal cotton-holder) delegitimates the meeting on this basis, arguing that those most in need of the patronage opportunities which the *gram sabha* represented were unable to attend due to their need to engage in daily wage labour.

Despite the presence of a female *sarpanch* in both Bantala and Nandanapuram, it was the *sarpanch*'s husband, Pallav, who was clearly dominant in Bantala. Although elected in a year when the post was reserved for BC females, the *sarpanch* was from the dominant BC Kuruma caste, and her appointment allowed her husband to consolidate his already considerable position of power in the village. The situation in Bantala suggests the way the reservations system is mediated through the power structure arising from the land-holding, caste and gender profile of each village, and the operation of reservations depends upon how this pre-existing power structure is legitimated.

The research illustrates the significance of the *sarpanch* in allowing participants to access patronage to facilitate their negotiation of risk. This was evident in the case of Nipa (female, SC Madiga, landless) in Bantala who claimed that she was not allocated a house as part of the Indiramma scheme because the *sarpanch* had not written a request for her. Similarly, Ranjan (male, BC Chakali, landless) in Nandanapuram stated that the *sarpanch* had stopped distributing his pension to him because of a dispute.

The research also highlights that in Orgampalle, in particular, there is a tendency to allocate the responsibility for legitimation of the PRI structure to village elders. The intersection of a traditionally ingrained hierarchy with a democratic practice charged with resource allocation lends itself to a certain legitimation of corruption within political practice. This is highlighted in the discourse of Amita (female, BC Chakali, smallholder), a widow in Orgampalle, in her assertion of politicians as 'important people' who are entitled to secure greater access to resources for themselves because of their privileged positions.

The legitimation of corruption in political practice expressed by Amita suggests the influence of Telangana's imperial past on current views regarding democratic legitimacy in the region. Here, politicians have replaced rulers as dispensers of patronage, and a certain arbitrariness in the allocation of resources is excused as an almost inevitable aspect of the hierarchy which the political system is seen to represent. This attempt to blend the ongoing influence of hierarchy in the context of Telangana with a democratic ideal which emphasises equality and justice in resource allocation lends itself to the particular 'vernacularisation' (Michelutti 2007: 639) associated with the legitimation of democracy in the region.

The analysis indicates the way the caste composition of particular villages also impacts upon the electoral process. In Nandanapuram, despite their numerical majority, SC Madigas rely on FC Reddys to secure agricultural wage labour. According to participants, this relation of dependency allows the FC Reddys to manipulate the voting behaviour of the SC Madigas. While the 'free' wage relation (as opposed to a situation of bonded labour) and the power associated with their numerical presence and landholding in Nandanapuram allows SC Madigas to mobilise in order to demand higher labour rates in Nandanapuram, it does not overcome the pressure exerted through the more concentrated land ownership and wider political power associated with the FC Reddys in the village.

There is, however, also a discourse of rights expressed by a number of participants with regard to democratic practice. This is notable in the case of Sudeep (male, SC Mala, marginal cotton cultivator) in Bantala whose assertion of his right to justification for the cotton prices he is offered is linked to his living in a democratic society. Sudeep is, however, debt-free. He also has a lucrative income as an auto-rickshaw owner which means his subsistence is relatively independent of village power relations.[4] This is not the case for the more vulnerable in the village who, as highlighted, rely on the powerful for patronage as the means to negotiating risk. In their case, the assertion of their right to justification carries the danger of potentially exacerbating their exposure to risk.

While villagers in Bantala were active in wider mobilisations, this political engagement was orchestrated by Pallav. Despite the concerns of all participants, including the powerful, regarding Bt cotton, there is no engagement with an NGO for agricultural purposes. This may be related to the fact that alignment with an NGO would involve a sharing of power through an affiliation beyond the village at risk to village autonomy. Given the concentration of power in Bantala, this may not be welcomed by power-holders due to the potential threat to their position and the heightened risk of challenge to the oppressive power structure in the village that such an arrangement might entail.

In both Orgampalle and Nandanapuram, NGOs exert considerable influence. This is particularly true of Orgampalle where my research trips to the village had to be agreed in advance with Crops Jangaon. In order to promote the alternative practice of organic farming in the village, Pradnesh has been obliged to share his power with this NGO. The arrangement is likely to have been made easier by the strong cohesiveness of Orgampalle arising from its small size and caste homogeneity, as well as the fact that Pradnesh already shares power with the *sarpanch* in a neighbouring village. In addition, the perceived legitimacy of Pradnesh's position in Orgampalle is far more secure than is the case for the power-holders in Bantala where the two Scheduled Caste cultivators (Ashna and Sudeep) are strongly aware of their oppression.

The strict control maintained over my visits to Orgampalle by Crops Jangaon, as well as my need to request time alone with the villagers on my initial research trip, highlighted the extent of the NGO's intervention in the village. Nonetheless, in both Orgampalle and Nandanapuram, interaction with NGOs serves a number of functions. It not only provides cultivators with options other than Bt cotton through assisting with the supply of non-Bt seeds and providing agricultural extension services for alternative cultivation methods; NGO engagement also serves to politicise village society through building inter-village awareness of risk as a collective, rather than an individual, concern. The latter may well be where the uncertainty concerning interaction with NGOs among power-holders in Bantala lies given that the village power structure currently serves the collective interests of the powerful but not necessarily those of the Scheduled Castes and landless. Both of these NGOs also challenge the state to support more regionally specific solutions to agrarian risk than the current situation with Bt cotton allows.

The protests against Bt technology, while not fully inclusive, extend awareness of the right to justification in ways which heighten the concern with legitimacy in relation to power structures and access to resources. Although it is male Backward Castes with a variety of land-holdings who are actively campaigning against the technology in the current study, it has been highlighted that attempts at mobilising Scheduled Caste females in protests have also occurred in the region. This suggests that Bt cotton protests are spreading awareness of the right to justification for exposure to risk to wider (although by no means all) groups of the vulnerable.

The analysis also indicates the centrality of the issue of Bt technology to perceptions of state legitimacy in the villages. The questionnaire results in Fig. 7.1 highlight that the legitimation of Bt cotton, and a depoliticisation of the anxiety with regard to its adoption, coincides with a greater sense of recognition and perception of the legitimacy of the state government in Bantala. The delegitimation of Bt cotton, however, coincides with a sense of misrecognition by both the state and Indian governments among participants in Nandanapuram given the perceived failure of both to respond effectively to their protests and exposure to risk. This misrecognition is associated with an adverse impact on the perceived legitimacy of both governments. Despite the strong opposition to Bt technology in Orgampalle, the sense of misrecognition was alleviated somewhat as a result of the consultation process undertaken by the then Minister of the Environment and Forests, Jairam Ramesh, in relation to Bt brinjal.

The study also indicates that, in Bantala, a village characterised by its inequality and concentrated power structure, there is a greater perception of equality facilitated by the democratic process (Fig. 7.2). This is particularly asserted by the more powerful Forward and dominant castes. Ashna (female, SC Madiga, marginal cultivator), however, describes the suppression of lower castes in the village and the marginalisation of perspectives with which it is associated. This highlights the way in which legitimation struggles often entail differences in perspective between those who are currently benefiting from an unequal access to resources and not at immediate risk and those who are significantly exposed to risk as a result of such inequity.

In Nandanapuram, views are more diffuse with regard to the contribution of democratic practice to bringing about greater equality. Participants refer to the corruption in the political process, as well as the low education levels of the electorate. They also assert their dissatisfaction with the absence of a single authority which would be willing to take the decision to ban Bt technology. In Orgampalle, participants argue that the wealthy, particularly multinational organisations, are being favoured by the Indian government in decision-making on Bt technology. The analysis suggests, therefore, that the legitimacy of the new state of Telangana will be inextricably linked to the way in which the issue of Bt technology, and the agrarian crisis more generally, is negotiated by the state government.

Forrester (1970: 21) claims that 'the tragedy of Telangana is that so little was done to identify or deal with the legitimate grievances of the area for so many years.' The research highlights the way in which this historic lack of recognition and representation influences the contemporary legitimation of democracy in terms of securing justice in redistribution.

The analysis indicates the way the idea of democracy is translated into a political practice which is assessed for its legitimacy on the basis of how power is exercised and resources, and risk, distributed through it. As noted, this imperial past contributes to a certain legitimation of corruption associated with patronage and the culture of 'money politics' described by Powis (2003: 2620). This illustrates the view of Abromeit and Stoiber (2007: 35) that understandings of legitimacy 'vary with the historical and societal context.'

Rao and Sanyal (2010: 146) claim that, through *gram sabhas*, 'villagers are constitutionally empowered to make decisions regarding budgetary allocations for village development and beneficiary selection for antipoverty programmes.' These authors also assert the deliberative dimension of the meeting, arguing that it is, in many ways, a public sphere which should be free 'of the distorting effect of inequalities or coercion' (Rao and Sanyal 2010: 146).

The analysis highlights, however, that the process of legitimation produces a particular 'vernacularisation of democracy' (Michelutti 2007) associated with the exercise of power arising from the particular composition and power structure of each village. It also indicates how the *gram sabha* is linked to the accessing of opportunities for patronage to facilitate the negotiation of risk. This supports Kumar's (2006: 236) finding that villagers generally attend *gram sabha* meetings 'demanding or expecting some personal benefit.'

Ranjan's (male, BC Chakali, landless) non-attendance at the *gram sabha* following his dispute with the *sarpanch* in Nandanapuram, a disagreement which has led to the suspension of his pension, suggests the further finding by Kumar (2006: 229) that 'only cronies of the *sarpanch* attended the *[gram] sabha*.' The ability of the *sarpanch* to control patronage opportunities in this way serves to reinforce the power structure, given that challenges to the power of the *sarpanch* are likely to have a detrimental impact on the ability to negotiate risk. This is noted by Omvedt (1993: 291) who argues that the 'dispensing of patronage function[s] mainly to demobilise the oppressed.'

The centrality of the *sarpanch* position within village power relations is noted by Powis (2003: 2618) who claims that '[p]olitical competition is most intense around *sarpanch* elections.' The situation in Bantala with regard to reservations for females for this position suggests the view of Corbridge and Harriss (2000: 209) that the election of women [in *panchayat* or council elections] 'has only rarely threatened the *de facto* power which male relatives exercise over the political process.' The current analysis indicates that reservations will have a different outcome depending upon the composition and exercise of power in particular villages. It also suggests that their success in mobilising women relies upon the extent to which the engagement of females in politics is legitimated in the villages by females themselves.

As in the case of the current study, Kumar's (2006: 209) research highlights the highly gendered nature of the *gram sabha*. This has led a number of the more vulnerable females in the current study to prioritise attendance at Self-Help Group meetings as their preferred means of risk negotiation. As Galab and Rao (2003: 1281) note, '[p]articipation in SHGs has improved the access of women to credit.' Because SHGs are defined by personal relations between females often of the same caste, they represent a lower cost, lower-risk source of debt. The analysis suggests, however, that given their prioritisation by females, SHGs may also contribute to the ongoing poor representation of women within institutionalised democratic practice in the villages and expose the already extremely vulnerable to the additional risk of indebtedness.

The research highlights that the traditional power structure associated with village elders also remains central. The fact that it is the elders in Bantala who are purported to be planning a visit to Orgampalle to assess the viability of organic cultivation indicates their ongoing role as representatives of the village in relation to the negotiation of risk, a finding also noted by Pur and Moore (2010: 619).

The assertion by Sudeep (male, SC Mala, marginal cotton-holder) that 'since we are living in a democratic country, this has given us the right to ask for our rights....Now I have enough freedom to go and ask what rate I'm getting [for my cotton] and why I'm getting it' explicitly illustrates the connection between the legitimation of risk and democracy which this book has sought to highlight. It is noted that the ability to claim recognition for rights is strongly mediated through power relations (in Sudeep's case, as a cotton cultivator, the negotiation of the power relation associated with cotton traders has been identified as significant); nonetheless, the idea of democracy suggests to Sudeep that cotton cultivators have a right to a fair and justifiable price for their cotton. The emergence of this idea and its application to everyday life often lends itself to the development of another, far more radical one—that the power relation through which the cotton price is determined is itself open to challenge on the grounds of its legitimacy.

Sudeep's statement highlights Forst's (2014: 6) claim that 'the basic question of justice is not what you have but how you are treated.' However, Forst's assertion fails to fully appreciate how access to resources defines one's place in a power structure in ways which impact upon the freedom to assert recognition for one's right to justification. Thus, while the vulnerable may well be treated unjustly, the recognition of this is often reliant upon their representation by those who are less immediately at risk. In this regard, Forst's (ibid.: ix) view that 'the first question of justice is power' is insightful. However, in relation to the legitimation of risk and democracy and the demand for the right to justification within this, it is not so much power itself, but how it is comprised, exercised and negotiated which becomes more specifically of significance.

The impact of the power structure on the potential for 'epistemic injustice' (Fricker 2007) is also key. In the current research, this is indicated in relation to the animal deaths in Bantala which participants in all three villages perceived as being linked to Bt cotton. A number of villagers in both Nandanapuram and Orgampalle expressed the hindrance associated with their lack of a scientific education on their ability to develop their own knowledge or assert their right to justification through the political process to gain clarity in this regard.

The situation with regard to the animal deaths highlights the 'evidence trap' which Renn (2008: 133) argues risk represents. This means that, as Beck (1995: 131) also notes, the 'burdens of damage and proof [of risk are] foisted on injured parties.' In the case of Bt cotton, the cultivators who are potentially being placed at risk through its adoption are also being misrecognised by scientists who have come to investigate the deaths. Herring's (2008: 155) view that the linking of animal deaths to Bt cotton is a fabrication to serve the interests of NGOs was belied by an interview with the village vet who had personally witnessed the 15 dead animals in Bantala. If these deaths are not linked to Bt cotton, then the villagers should have been informed of the reasons for them. Instead, they have had no feedback from investigators who came to the village in this regard—an illegitimate misrecognition of their right to justification in relation to their practice of Bt cotton cultivation and an epistemic injustice which serves to contribute to their sense of risk. Given the absence of feedback and subsequent official advice that they should not allow their animals to graze on Bt cotton, this sense of risk is entirely understandable.

Legitimation, Power and the State of Telangana

As part if its analysis of the legitimation and delegitimation of Bt cotton among cultivators, this book explored the significance of Bt technology to the state's attempts to secure its legitimacy in pre-secession Andhra Pradesh. It argues that this will be an ongoing concern for the state of Telangana and that the new government will be obliged to negotiate many competing influences in this regard. These include powerful forces not only from within Telangana itself in the form of cultivator mobilisations, a strong NGO sector, and radical Naxalites, but also the central government in Delhi, the World Trade Organisation and World Bank, as well as multinational corporations, such as Monsanto.

The analysis illustrates the way in which the unrest associated with Bt technology in Telangana is embedded within an ongoing struggle related to the demand for justification with regard to the exercise of power and the differentiated access to resources in the region. This historical concern meant that the pre-secession state was required to balance the demands asserted by civil society protests on its periphery with those of the wider power structure within which the issue of Bt technology is embedded. This involved the pre-secession state in negotiating the threat to its legitimacy on a number of different levels—local, national and global.

Since Narendra Modi's right-wing Bharatiya Janata Party (BJP) rose to power in Delhi in 2014, concerns are ongoing that the central government's emphasis on privatisation and liberalisation is clearing a path for Monsanto to push ahead with field trials in the absence of a clear regulatory mechanism.[5] This has seen the new state of Telangana looked set to join the Punjab, Haryana, Delhi, Maharashtra and the divided Andhra Pradesh in issuing 'No Objection Certificates' for trials of Bt food crops, including not only brinjal but also potato, groundnut and chickpea.[6]

The analysis of the villages highlights the significant potential for a 'legitimation crisis' (Habermas [1973], 1976) in the region which such an approach represents. While participants in Bantala strongly legitimated Bt cotton despite their concerns, the opposition to Bt brinjal was asserted by participants in all three villages. Anxieties in this regard related to fears concerning animal deaths which, despite their dismissal by academics such as Ronald Herring, remain unresolved. If, as is believed by the former Director for Animal Husbandry in Warangal and the village vet for Bantala, these deaths were caused by pesticide residues rather than the Bt plant itself, then this is still a significant cause for alarm with regard to the introduction of Bt food crops.

It is clear that the development plans of the state government have a significant impact upon risk negotiation in the villages. The policy document issued by the new government's Planning Department in 2016 entitled 'Reinventing Telangana' outlined the government's commitment to building a *Bangaru* (Golden) Telangana.[7] The report highlights the government's intention to:

- repair and revive tanks throughout the region to ease the pressure on ground-water (Mission Kakatiya) (p. 5);
- accelerate employment in the services and industrial sectors (p. 5);
- target welfare programmes to the poor (p. 6);
- restore faith in government hospitals (p. 6);
- provide dowry assistance for 'economically backward' Scheduled Castes and Scheduled Tribes (p. 6);
- provide drinking water for all (p. 6);
- improve the quality of education (p. 110)
- address the causes of land degradation (p. 115).

As the analysis highlights, these are key steps which are long overdue. The question of irrigation is particularly important in a dryland area, such as Telangana. Rao (2014: 12), too, notes the urgent need for the 'rejuvenation of tanks' to eliminate the requirement for expensive borewells. The latter exacerbate the risk of water shortage for the collective through lowering ground-water levels, as well as heightening exposure to debt for individual cultivators. As this book has sought to illustrate, policies such as those above are essential not only as a concern for social justice and to address the ongoing and significant issue of farmer suicides in Telangana; they are also crucial to securing the legitimacy of the new state itself.

The report also announces the introduction of a 2BHK housing scheme with the aim of providing two-bedroomed homes for the homeless (p. 7) and intends to promote participation in Self-Help Groups (SHGs) in the region (p. 111). The experience of the Indiramma initiative and its contribution to indebtedness illustrated in the current study suggests the need for care with regard to the housing scheme to ensure that it does not promote the tendency to Sanskritisation in ways which further add to indebtedness. This involves ensuring that the funding provided is adequate and that it is carefully targeted given the impact of village power structures on how welfare schemes are distributed to the vulnerable.

Because of the numbers of female suicides in Telangana and the fact that SHGs were found to be a key source of lower-interest borrowing for the more vulnerable Scheduled Caste females in the current study, the focus on SHGs in the policy document is important. Again, however, there needs to be careful attention to the way the promotion of SHGs is implemented. The situation of Salma (female, SC Madiga, landless) in Nandanapuram highlights that it is important to ensure that the access to credit supported by SHGs does not itself contribute to exposing already high-risk females to the additional risk of indebtedness—admittedly, through a lower-interest, more collectivised, form of borrowing but still resulting in indebtedness which must be paid back from meagre incomes nonetheless.

The experience of Orgampalle and the mitigation of risk associated with lower cost cultivation methods suggest that, given the ongoing issue of farmer suicides, the government needs to focus on reducing costs and cultivator indebtedness through encouraging and supporting a more collective approach to agriculture. The analysis indicates that new ways to minimise the costs of labour charges for cultivators which do not impact upon the livelihoods of the landless should also be sought through, for instance, a system of state subsidies for agricultural labour. Such subsidies should seek to minimise the potential for labour costs to be elevated as a result of village power structures.

The degree to which the state is perceived to be performing its key function of offering citizens protection in risk society is central to its ability to secure its own legitimacy. The risk to state legitimacy which the issue of Bt technology represents in Telangana is particularly significant given the history of volatility in the region and the long struggle of its people to secure a more legitimate exercise of power in the allocation of resources and the negotiation of risk.

As the analysis of the villages highlights, the new government has a number of options in relation to Bt technology. It can adopt an authoritarian approach where Bt food trials are pushed through by power-holders in ways which seek to bypass a formal legitimation process. This relates to the 'quasi-totalitarian moments' (Fraser 2008: 140) in democratic societies in which the power structure itself drives the legitimation process and attempts to override the right to justification of those who are placed at risk as a result.

Such an authoritarian approach is redolent of the strategy of risk negotiation adopted by power-holders in Bantala. As highlighted, such a tactic can only be secured through the promotion of the perception of a generally enhanced prosperity in ways which secure legitimation for the ongoing suppression of the most vulnerable. This strategy comes at a cost not only to democratic legitimacy but also to epistemic justice given the ongoing fears for the future sustainability and safety of Bt technology which have not been adequately addressed. If these fears are more firmly actualised, they will represent a significant threat to the state's legitimacy given the primary basis on which such legitimacy is founded is that the state serve as the ultimate protector of its citizens.

The process of legitimation seeks to challenge such totalitarian moments in democratic societies and to assert the right to justification of the people as a concern for justice. Given the volatility, vulnerability and history of the context of Telangana, the risk of legitimation crisis associated with a totalitarian approach is acute. Many of those in Telangana are highly politicised as a result of the historic struggle in the region, and the majority of citizens are extremely aware of the links between their ability to negotiate risk and their demands for democratic legitimacy. As the response in the region to the treatment of Naxalites by the pre-secession government highlighted, an authoritarian approach which seeks to override the process of reason-giving and justification in the struggle for legitimation may result in an early loss of legitimacy which the new state may have difficulty in regaining.

The current analysis illustrates the legitimation struggle that the issue of Bt cotton represents across the villages. This relates to whether the technology's unreliable economic benefits in a high-risk context merit the additional economic and environmental risks which are perceived by many in terms of its adoption. In Nandanapuram, this struggle is evident within the village itself. The situation in the village highlights that the widespread adoption of Bt cotton cannot be assumed to indicate either its unproblematic legitimation or that significant opposition to Bt food crops does not exist.

The importance of the presence of radical Naxalites whose legitimacy is inversely related to that of the state, as well as the strong NGO sector in Telangana, cannot be under-estimated. The impact of their assertions of the ongoing inequality and risk exposure in Telangana was, as the analysis has highlighted, significant within the struggle for legitimacy of the pre-secession state.

The validity of the risk discourse asserted by NGOs is reinforced by the fact that Telangana was ranked second in India in 2014 with regard to farmer suicides (Goyal 2015: 267). The fact that these suicides are increasingly occurring among cultivators with holdings of up to 25 acres (ibid.: 283) suggests the growing problem of the non-viability of agriculture. It also reinforces the need for a focus on reducing cultivation costs and indebtedness as the key to alleviating agrarian risk. The expanding problem of farmer suicides represents a significant challenge to the legitimacy of an approach to agrarian risk which involves broadening the application of a technology which, as the current study highlights, has the potential to exacerbate, as well as alleviate, cultivator risk, indebtedness and, given their links to such indebtedness, potentially suicides.

The analysis of the villages highlights that the risks associated with Bt technology have far more validity than Herring (2008: 146) implies in his view of self-interested NGOs and cultivators fabricating stories about animal deaths and crop failure in order to further their own interests. Instead, the research indicates the way in which the legitimation struggle which Bt technology represents is embedded within a far wider conflict related to the negotiation of risk more generally. This is associated with a critical focus on the way in which power is exercised and resources allocated as a concern for justice in the villages and beyond. The struggle is differently translated in accordance with the power constellations of particular contexts, but the resulting contestation and differences of perspective are all part of the same struggle for the legitimation of risk and democracy.

The new state will increasingly be drawn into conflicts between the demands of a generally politicised population characterised by a high degree of inequality and poverty, and the complex power negotiations involved in being part of both a larger Indian state and an increasingly interconnected world. In terms of securing the state's legitimacy within this, the village analysis offers a number of key insights.

The liberalisation of the pre-secession state, and the adoption of Bt cotton within this, contributed to 'agricultural individualisation' (Vasavi 2010: 77–80) as a response to agrarian risk. This led to cultivators accessing credit and inputs as atomised individuals. Many public institutions which had focussed on encouraging collective behaviours were eroded. These included state co-operatives such as the Andhra Pradesh Irrigation Development Corporation, the Andhra Pradesh Seeds Development Corporation and the Co-operative Spinning Mills (Galab et al. 2009: 190).

The pervasiveness of indebtedness suggests that initiatives aimed at reducing cultivator costs through subsidies and the reinstatement of co-operatives may well be a more effective means to addressing agrarian risk than the introduction of additional GM crops. The analysis of Orgampalle highlights the benefits associated with encouraging solidarity and a more collective response to risk, particularly in relation to reducing the costs of cultivation, and the sharing of key resources, such as oxen, water and labour. The research also suggests the urgent need for debt levels to be measured along with economic growth, and for such a monitor to be taken into account when exploring the viability of Bt technology relative to alternative strategies such as organic and NPM cultivation, given the particular vagaries associated with dryland agriculture in India.

The reduction in government investment on extension services, and their increased privatisation as part of the neoliberal reforms in the pre-secession state (Galab et al. 2009: 189) reduced the social supports available to farmers in their attempts to negotiate risk. As Rao (2014: 10) emphasises, this needs to be reversed given that '[a] new social framework which is participatory and accountable to stakeholders is a prerequisite for inclusive and sustainable development in the new state.'

The government should also draw upon the knowledge and reach of the extensive NGO network in Telangana with regard to its attempts to alleviate agrarian risk.[8] This is also noted by Bhagwati (1998: 39) who argues for 'the need for government and NGOs to work together on an anti-poverty agenda that would begin with efficiency and growth.' This collaboration would serve as a check on the legitimacy of the exercise of power by both, provided that NGOs maintain their autonomy and do not become co-opted. It has been highlighted that NGOs are extremely diverse in their orientation to the state and to cultivators. However, despite the issues which have been highlighted with regard to their somewhat overbearing intervention in Orgampalle, the NGOs involved in the current study contribute to the choices which are available to cultivators and offer valuable extension services. This work needs to be supported through government investment with the aim of developing the ability of cultivators to alleviate their risk exposure in ways which allow them to act upon their own understandings of agrarian risk. This involves finding ways to secure the supply and availability of non-Bt cotton seeds to facilitate cultivation options other than Bt cotton adoption. As highlighted by Ramanjaneyulu and Kuruganti (2006: 563), the government should make 'more effort to study all other options available before making any categorical statements on [Bt] technology.'

Most importantly, this analysis has highlighted the need for the new government to develop a measure for indebtedness and to ensure that this is monitored along with statistics on economic growth. The concern with indebtedness needs to inform considerations of whether the ongoing strategy of encouraging access to credit, and supporting a high-cost cultivation method while presiding over the exclusion of others, is really the most responsible method for facilitating the risk negotiation of vulnerable cultivators in the region.

Beetham (2013: 121) highlights that 'the state is responsible for determining the rules which govern all other power relations in society…. Whatever the limits in practice to this power, its possession makes the state the site of intense struggle to control it, or to influence those who do.' It is clear from the analysis that, as part of this, the new government will need to address the power relations in the region which significantly exacerbate agrarian risk and which secure its uneven distribution.

The impact of local power relations on risk negotiation and differentiation could be challenged through the state-sponsored co-operatives discussed earlier. These should focus on the sharing of key resources and the bulk ordering of inputs to facilitate lower prices for cultivators and organised across villages along dimensions of differentiation such as caste and gender. Such inter-village arrangements would seek to overcome the constraints associated with the village power structure with regard to the unequal access to resources. The collective approach to risk would also challenge the power nexus of middlemen and traders in the agricultural context given that their lack of regulation is recognised as contributing to higher cultivation costs, the restriction of farmer choice and the exacerbation of the risks of indebtedness (Varshney 1998: 98; Frankel 2005: 261).[9]

Walker and Ryan (1990: 355) note the need for a 'decentralised regional approach' as the key to alleviating poverty and risk in dryland agriculture. The policy document of the Telangana government (endnote 7) highlights the launch of the *Gram Jyothi* programme to strengthen the *panchayats*. As the current research highlights, however, care will need to be taken to ensure that further initiatives in this area do not simply result in enhancing the power of the already dominant.

The analysis highlights that greater attention needs to be paid to village power relations to ensure that government initiatives are effectively implemented. This requires a micro-understanding of the villages in terms of how the legitimation of new initiatives can be secured either through or around power structures. Such a nuanced understanding can only emerge through spending time in the villages with both power-holders and the vulnerable in order to gain a sense of the way in which power is exercised and risk exposure differentiated in particular villages. This approach would permit the emergence of creative means to ensure that the delivery of development resources does not serve simply to reinforce existing power relations. The analysis suggests that this is currently the perception of the operation of the PRIs, a factor which is contributing to their delegitimation.

The implementation of development initiatives may involve, for instance, working with power-holders directly to persuade them of the benefits to their legitimacy of a new approach. Or it might entail regular visits from representatives of the Mandal Parishad to the villages to implement welfare schemes, rather than simply relying upon an often poorly attended *gram sabha* as the means to doing so. The recent change in the administrative structure of Telangana and its division into smaller units (see Map 1) should allow the government to be far more active in terms of its direct engagement with villages. This involvement should aim to ensure that welfare programmes are informed by the needs of the vulnerable, and are accessed by them, in ways which negotiate the village power structure more skilfully than the institutionalised *gram sabha* currently permits.

Rao (2014: 11) argues that 'a smaller state being more easily accessible to the common people can intelligently and speedily grapple with its problems.' The analysis highlights that an in-depth understanding of the context, and its power relations, is the key to achieving the type of recognition and representation which makes both redistribution and democratic legitimacy possible in the negotiation of risk. However, the state will also be subject to 'the dilemma of democracy' referred to by Young. Young (2000: 228) notes that, while citizens have a greater influence over decisions in smaller political units, these units themselves have little power 'to influence far-reaching relations and actions that fundamentally affect their local conditions' (ibid.). The legitimacy of the new government will, therefore, rely upon its ability to come to terms with power relations, not only at the micro-level of the villages but also at the levels of the Indian state and global power-holders.

In terms of the wider power structures associated with Bt technology, Raina (2006: 1624) notes that the 2006 Indo-US Knowledge Initiative significantly strengthened the influence of US multinational corporations within India's attempts to define its own knowledge construction with regard to its negotiation of risk. As was discussed in Chap. 3, the pre-secession state was involved in multi-level negotiations, including those with multinational corporations, such as Monsanto, as well as the central government in Delhi, given concerns for its own legitimacy arising from the protests of its citizens against Bt technology. The new government of Telangana will be required to enter into the same negotiations and walk the same tightrope of trying to balance competing validity claims which the issue of Bt technology involves.

Within the negotiation of its legitimacy in relation to Bt technology, the new government should take note of the view of Ayyangar (2003:

4429) that '[t]he closer the powerful get to the people, the more the people feel that their voices can be heard.' The benefits to state legitimacy associated with such an approach were highlighted in the analysis given the contribution of Jairam Ramesh's efforts at consultation to the enhanced perception of recognition and state legitimacy in Orgampalle, despite their delegitimation of Bt technology.

Habermas' (2008: 330) argument that '[a state] cannot preserve the necessary level of legitimacy in the long run unless a functioning economy fulfills the preconditions for an acceptable pattern of distribution' is also noted. This relates to the wealth creation and increases in productivity to which Bt technology proponents argue the technology itself has contributed. However, as Beetham (2013: 140) emphasises, 'the actual relationship between economic performance and government legitimacy is a complex one, depending upon the pattern of distribution of economic costs and benefits, as well as on the overall level of performance.'

It is, therefore, not sufficient to create the conditions for economic growth in poor contexts in order to secure political legitimacy; consideration must also be given to the way in which opportunities for wealth creation are distributed among citizens, as well as any risks with which such opportunities may themselves be associated, as a concern for justification and justice.

The research suggests that is on the three dimensions of justice—redistribution, recognition and representation—and the concern with the right to justification which informs them that the new government will need to focus if it is to secure its legitimacy. This is particularly crucial for Telangana given the long struggle for legitimation in this highly volatile and vulnerable region of a world at risk.

Notes

1. A significant barrier to NPM adoption is the highly specialised knowledge and training which it involves. Given Nand's presence in Nandanapuram, as well as the involvement of the Deccan Development Society, this is not an issue in Nandanapuram.
2. Though it is, of course, also noted that the power structure associated with village elders must itself be made subject to a process of legitimation.

3. The specific impact of crop loss of Bt cotton at an individual level can be found in the wider study entitled *The Legitimation of Risk and Democracy: A Case Study of Andhra Pradesh, India*. This is available online at: https://cora.ucc.ie/handle/10468/1688. pp. 386–388. The consequences of the additional costs of Bt cotton on cultivator risk exposure in poor seasons could be seen in the case of Ambu (female, SC Madiga, cotton small-holder) in Nandanapuram. In the far from atypically catastrophic 2010/2011 season in which the research was undertaken, Ambu lost her crop on two flooded acres and managed to obtain a low yield from her third. This was after an expenditure of almost Rs 22,800 (Rs 7600 per acre) on seeds, pesticides and fertilisers. This meant that she made a total loss of Rs 49,000 in the 2010/2011 season. (Her other most significant cost was labour at Rs 11,900 per acre). Aruni (female, dominant caste, marginal cotton cultivator) in Orgampalle obtained a similar yield overall (and a lower price for her cotton) but still managed to make a small profit (Rs 5100), given the absence of spending on inputs, as well as lower labour costs and the absence of costs for tractor/oxen hire. These two vulnerable female cultivators were, therefore, in markedly different positions when facing the coming season.

4. This auto-rickshaw was itself purchased through borrowing, now paid back in full, highlighting that indebtedness, like Bt cotton, can be risk-alleviating, as well as risk-enhancing.

5. See http://www.gmwatch.org/index.php/news/archive/2015-articles/16038. Accessed on 2/4/2017.

6. Available at: http://articles.economictimes.indiatimes.com/2015-11-13/news/68252428_1_field-trials-gm-crops-genetic-engineering-appraisal-committee. Accessed on 2/4/2017.

7. This is available at www.telangana.gov.in/PDFDocuments/Socio-Economic-Outlook-2016.pdf. Accessed on 31/3/2017.

8. While the policy document (see endnote 7) notes the intention to involve NGOs with regard to female empowerment, the role of NGOs could have a much broader utilisation given their significant locally-specific knowledge in the area of cultivation. As has been highlighted, care would need to be taken with how such NGO involvement was managed given that village power-holders may perceive NGO intervention as a threat to their own positions. NGOs may also themselves adopt a domineering approach to villagers in ways which contribute to epistemic injustice and agricultural deskilling and deny the knowledge of farmers themselves.

9. As Varshney (1998: 98) highlights, the idea of regulation is likely to lead to political challenges from the central government as the BJP party is one of the few political parties to support the cause of traders and middlemen.

Bibliography

Abromeit, H., & Stoiber, M. (2007). Criteria of democratic legitimacy. In A. Hurrelmann, S. Schneider, & J. Steffek (Eds.), *Legitimacy in an age of global politics* (pp. 35–56). New York: Palgrave Macmillan.

Ayyangar, S. (2003). Janmabhoomi meetings in two villages. *Economic and Political Weekly, 38*(42), 4426–4429.

Beck, U. (1992). *Risk society: Towards a new modernity.* London: Sage.

Beck, U. (1995). *Ecological politics in an age of risk.* Cambridge: Polity Press.

Beetham, D. (2013). *The legitimation of power.* London: Palgrave Macmillan.

Bhagwati, J. (1998). The design of Indian development. In I. J. Ahluwalia & I. M. D. Little (Eds.), *India's economic reforms and development: Essays for Manmohan Singh* (pp. 23–39). Delhi: Oxford University Press.

Binz, C., Harris-Lovett, S., Kiparsky, M., Sedlak, D. L., & Truffer, B. (2016). The thorny road to technology legitimation – Institutional work for potable water reuse in California. *Technological Forecasting and Social Change, 103*, 249–263.

Cooke, B. (2001). The social psychological limits of participation. In B. Cooke & U. Kothari (Eds.), *Participation: The new tyranny?* (pp. 102–121). London: Zed Books.

Corbridge, S., & Harriss, J. (2000). *Reinventing India: Liberalization, Hindu nationalism and popular democracy.* Cambridge: Polity Press.

Forrester, D. B. (1970). Subregionalism in India: The case of Telangana. *Pacific Affairs, 43*(1), 5–21.

Forst, R. (2014). *Justice, democracy and the right to justification.* London: Bloomsbury.

Frankel, F. R. (2005). *India's political economy 1947–2004.* New Delhi: Oxford University Press.

Fraser, N. (2008). *Scales of justice: Reimagining political space in a globalizing world.* Cambridge: Polity Press.

Fricker, M. (2007). *Epistemic injustice: Power & the ethics of knowing.* Oxford: Oxford University Press.

Galab, S., & Rao, N. C. (2003). Women's self-help groups, poverty alleviation and empowerment. *Economic and Political Weekly, 38*(12/13), 1274–1283.

Galab, S., Revathi, E., & Reddy, P. P. (2009). Farmers' suicides and unfolding agrarian crisis in Andhra Pradesh. In D. N. Reddy & S. Mishra (Eds.), *Agrarian crisis in India* (pp. 164–198). New Delhi: Oxford University Press.

Goyal, L. C. (2015). *Accidental deaths and suicides in India, 2014.* Report from National Crime Records Bureau, New Delhi.

Gupta, S. P. (2002). *Report of the commission on India: Vision 2020.* New Delhi: Planning Commission, Government of India.

Habermas, J. (2008). *Between naturalism and religion.* Cambridge, UK: Polity Press.

Herring, R. J. (2008). Whose numbers count? Probing discrepant evidence on transgenic cotton in the Warangal district of India. *International Journal of Multiple Research Approaches, 2*(2), 145–159.

Kumar, G. (2006). *Local democracy in India: interpreting decentralization.* New Delhi: Sage.

Kumbamu, A. (2007). Discussion: Beyond agricultural deskilling and the spread of genetically modified cotton in Warangal. *Current Anthropology, 48*(6), 891–893.

Michelutti, L. (2007). The vernacularization of democracy: Political participation and popular politics in North India. *Journal of the Royal Anthropological Institute, 13*, 639–656.

Omvedt, G. (1993). *Reinventing revolution: New social movements and the socialist tradition in India.* London: M.E. Sharpe.

Powis, B. (2003). Grass roots politics and 'second wave of decentralisation' in Andhra Pradesh. *Economic and Political Weekly, 38*(26), 2617–2622.

Pur, K. A., & Moore, M. (2010). Ambiguous institutions: Traditional governance and local democracy in rural South India. *Journal of Development Studies, 46*(4), 603–623.

Raina, R. S. (2006). Indo-US knowledge initiative: Need for public debate. *Economic and Political Weekly, 41*(17), 1622–1624.

Rao, C. H. (2014). The new Telangana state: A perspective for inclusive and sustainable development. *Economic and Political Weekly, XLIX*(9), 10–13.

Rao, P. N., & Suri, K. C. (2006). Dimensions of agrarian distress in Andhra Pradesh. *Economic and Political Weekly, 41*(16), 1546–1552.

Rao, V., & Sanyal, P. (2010). Dignity through discourse: Poverty and the culture of deliberation in Indian village democracies. *The Annals of the American Academy, 629*, 146–172.

Renn, O. (2008). *Risk governance: Coping with uncertainty in a complex world.* London: Earthscan.

Srinivas, M. N. (1966). *Social change in modern India.* New Delhi: Orient Longman.

Stone, G. D. (2007). Agricultural deskilling and the spread of genetically modified cotton in Warangal. *Current Anthropology, 48*(1), 67–103.

Varshney, A. (1998). *Democracy, development and the countryside: Urban-rural struggles in India.* New York: Cambridge University Press.

Vasavi, A. R. (2010). Contextualising the agrarian suicides. In R. S. Deshpande & S. Arora (Eds.), *Agrarian crisis and farmer suicides* (pp. 70–85). New Delhi: Sage.

Walker, T. S., & Ryan, J. G. (1990). *Village and household economies in India's semi-arid tropics*. London: The Johns Hopkins University Press.

Young, I. M. (2000). *Inclusion and democracy*. New York: Oxford University Press.

9

Conclusion: Science, Power and the Struggle for Legitimation

GM Crops and State Legitimacy in Telangana

This book has illustrated the way in which the globally contested issue of GM crops is translated into a legitimation struggle involving three villages in the district of Warangal in Telangana. This conflict relates to ongoing concerns in the villages regarding the contribution of the crop itself and/or the cultivation practice which it involves to social, environmental and economic risk.

The analysis suggests that Bt cotton cannot be regarded as an adequate response to the agrarian risk of the context of Telangana. This is not only due to the anxiety among cultivators for its future sustainability as a cultivation practice but also given the particular indebtedness of Bt cotton cultivators when compared to those adopting the alternative low cost organic and NPM methods which this study has explored.

In terms of future research, it is recognised that the analysis of risk undertaken here can be extended beyond the focus on Bt cotton and that there are dimensions of risk negotiation which have not been covered. These include, for instance, a more thorough analysis of the impact of

© The Author(s) 2018
E.L. Desmond, *Legitimation in a World at Risk*,
https://doi.org/10.1007/978-981-10-6065-6_9

other *kharif* crops, or those cultivated during the second cropping season (the *rabi* season), on the wider risk negotiation of particular households. Likewise, an exploration of the way in which earnings from the National Rural Employment Guarantee Scheme (NREGS) have altered the power relation between agricultural labour and land-holders in the villages would also be of interest. Finally, as highlighted, Muslims and Scheduled Tribes represent two important minority groups whose exposure to risk is also severe but have not been included here. All of these aspects represent areas for inclusion in future studies either on Bt technology itself or as part of a wider analysis of risk negotiation in Telangana.

In his claim that Bt cotton in Warangal can be pro-poor, Herring (2008: 157) argues that '[f]armers of necessity count carefully; their numbers should count.' While this related to the calculation of yields and profits which cultivators noted Bt cotton had the potential to secure, this book argues that this concern with numbers should also extend to cultivator debt. It is vital that the impact of indebtedness is considered in any analysis of the way cultivators make decisions regardig their adoption of Bt cotton. Herring (ibid.: 156) also claims that it 'would be rash to uncritically accept farmer adoption as evidence for the efficacy of the technology.' This has been a central finding of the current study and is precisely why figures on cultivator indebtedness are key to the Bt cotton debate.

This book explored how Bt cotton's adoption, in conjunction with debt levels, was mediated through a concentrated power structure in Bantala. In the case of Nandanapuram, meanwhile, debt levels served to lock cultivators in to Bt cotton adoption given the increasingly desperate gamble to secure its potential for high yields. As the debt levels illustrate, however, this potential does not always materialise. This means that, as Gaurav and Mishra (2012: 25) note, 'Bt technology can be both risk increasing or risk decreasing.'

It is recognised that, despite its erratic performance in Telangana, Bt cotton is not the only reason for cultivator indebtedness and, in a good year, can contribute to an alleviation of risk; however, the comparison of the debt levels of Bt cotton cultivators with those of organic and NPM cultivators in the current study suggests the additional risk of indebtedness which Bt cotton cultivation, and its associated inputs, represents over time.

The acute and highly differentiated context of risk in Telangana is evident from the ongoing issue of farmer suicides. This book has sought to illustrate the complexity of the trade-offs involved in the decision to adopt Bt cotton given its potential for both risk and gain. This is also noted by Jakimow (2014: 409) who argues that commercial cultivation generally 'contains the possibility for both ruin and fortune' where the 'promises of a better, more progressive future are...possible...but by no means likely' (ibid.: 430).

It could be argued that, given the context of risk in Warangal and Telangana, the choice of any crop represents a risk. Being a commercial crop, cotton (whether Bt or not) is particularly high-risk given the fluctuating output prices and high costs for inputs, such as fertilisers and pesticides, especially in a context like Warangal which is prone to significant climatic vagaries and crop loss. This is also noted by Gaurav and Mishra (2012: 3) who argue: '[c]otton being a risky cash crop, varieties with [the] Bt gene are as susceptible to the risks in cotton cultivation as non Bt varieties are.' This relates to the inherent risks associated with dryland agriculture which Walker and Ryan (1990) also highlight.

The comparison of Bt cotton cultivators in Nandanapuram and Bantala with cotton cultivators in Orgampalle, however, suggests that cotton cultivation does not necessarily have to expose cultivators to extreme risk. The mitigation of its potential for risk relies upon the way in which power is exercised and agrarian risk negotiated as part of the wider context in which cotton cultivation is adopted. In Orgampalle, the diversification of crops (as can be seen from Table 6.1, only 17 per cent of village land is sown to cotton) and the collective response which focussed on minimising costs for all, not just the powerful, has ensured that cultivators are not indebted to the same extent as Bt cotton cultivators are. The significance of crop diversification for mitigating risk in dryland agriculture is noted by Gruère and Sengupta (2011: 334) and Walker and Ryan (1990: 243).

It could also be argued that, at least with Bt cotton, pesticide use has been reduced so costs are lower than previously and there is at least the *potential* for higher yields. These arguments, however, relate to an elapsed past and a hypothetical future. This book has sought to illustrate the present reality with which cultivators, particularly Bt cotton cultivators, are struggling given the constraints associated with the significant indebtedness of many and the anxiety which forms part of their lived experience.

The over-use and misuse of pesticides (Stone 2011: 393) in the past are being used as the basis for the legitimation of a technology which reduces the problem of pesticides but does not fundamentally address the problems of the farming practice associated with pesticides. This has seen, as Stone (ibid.: 391) highlights, the ongoing need for pesticides, in conjunction with Bt cotton, to combat other insects, such as sucking pests. The continued use of pesticides has contributed to ongoing concerns for soil erosion.

The situation in the villages highlights the 'individualisation of risk' noted by Beck (1992: 99) in risk society. Vasavi (2010: 77–80) claims that farmers approach the market as atomised individuals who must negotiate complex credit arrangements, as well as fluctuating prices of inputs and outputs, in a local context mediated through power relations and increasingly driven by impersonal globalised economic forces. Jodhka (2005: 22) claims that this 'autonomisation from the "traditional" rural economy and structures of patronage' has particularly isolated the lower castes as they face ongoing discrimination from upper castes.

As the analysis presented here suggests, however, the process of individualisation which Beck identifies in risk society is not complete. The construction of knowledge and risk, as well as the struggle for legitimation, are *social* processes. The individual is placed within a 'social risk position' (Beck 1992: 40) which he or she is obliged to negotiate not only as an individual but also as a social actor with agency. This has meant that risk society is giving rise to new forms of solidarity and co-operation which are emerging to challenge the attempts by market forces to diminish the social aspect of existence.

This re-assertion of the social in risk negotiation is evident throughout the research. The influence of conformist and prestige biases in Bt cotton adoption, the ongoing significance of borrowing from family and friends, the solidarity within caste communities, the existence of villages such as Orgampalle which seek to promote the idea of a collective, co-operative approach and, finally, the inter-village protests against Bt technology—all of these aspects serve to illustrate the way in which the individualisation of risk is being challenged through a collective response which asserts the social as the means to highlighting, negotiating, constructing and challenging risk.

Within the legitimation struggle associated with risk, the power relations of the social context and their impact upon the way in which risk is differentiated become of central importance. Beck (1995: 137) argues that 'wealth rises to the top while risks sink to the bottom.' This relates to the impact of class on risk society where the privileged access to resources of some alleviates their immediate exposure to risk at the expense of others. But, as Beck (ibid.) also notes, the structure of class conflict melts away in a world at risk and is 'recast in the heat of hazards.' This re-working of the differentiation associated with class can be seen in the nature of the mobilisations in contemporary risk society and the re-assertion of ideas of justice, justification and democracy in relation to the treatment of those made vulnerable through poverty. These ideas were evident in both the demands for secession in Telangana and the ongoing protests against Bt technology in the current study.

This book has highlighted that protests against Bt technology scale boundaries associated with the unit of the village, with the urban and rural, with individual states. This relates to the perceived need by many who are not currently immediately at risk to secure recognition and representation of the vulnerable as a concern for the long-term welfare of the collective. In this way, protests against Bt technology can be seen to have their foundations in class concerns (i.e. given the demand for recognition of the impact of an unequal access to resources on risk exposure) but also transcend them in order to focus more forcibly on the power relations which contribute to, and arise from such differentiated access.

In the 'new modernity' (Beck 1992) associated with risk society, the issue of power, and the legitimation of the way in which it is exercised, becomes of critical ontological significance. As Weber ([1968], 1978: 36) notes, these power relations are often derived from pre-capitalist conventions, such as those defined by tradition and affect. In risk society, norms arising from these previously unchallenged ideas are now being problematised given that they often serve as the basis for the differentiation associated with capitalist relations and the way in which resources are distributed in response to risk.

This book has highlighted the crucial role played by the state in influencing, through funding initiatives and legislation, the way in which risk is negotiated in the villages. This is also noted by Beck (1992: 202) who argues that, in risk society, the state is engaged in absorbing the 'social consequences [of technological change] and monitoring the risks.' But the state is also charged with defining a development model involving decisions on the relevance and likely contribution of particular approaches to the risk alleviation of its citizens. In a democratic state, such decision-making is directed through a legitimation process in which the perspectives of citizens themselves need to be recognised and represented.

In relation to Bt cotton in Telangana, the state is not only central to directing investment with regard to irrigation, extension services, subsidies, service and banking co-operatives and credit facilities; it also plays a key role in facilitating the choices which are available to cultivators—for instance, in terms of ensuring the availability of non-Bt seed varieties. As part of its concern for the protection of its citizens, the state also has a key role in regulating the middlemen who determine the prices of inputs, outputs and credit. The importance of the need for this regulative function in the pre-secession state of Andhra Pradesh was highlighted by Sridhar (2006: 1562).

The analysis suggests that the planned move by the new government in Telangana to begin trials with Bt food crops represents a significant concern, not only for state legitimacy but also for the potential contribution to risk globally which such a move would signify. While this proposed approval no doubt reflects attempts by the new state to negotiate the wider power struggle in which the issue is immersed, the Telangana government needs to be aware of the significant opposition to Bt food crops among many of its own citizens.

The resistance to Bt brinjal is highlighted in all three villages in the current research. The study suggests that, in the dryland context of Telangana, the priority needs to be to focus on offering extension services to facilitate decision-making at the level of the villages. This is given the complexity of the context and the precarious bases for the legitimation of Bt cotton in Bantala and Nandanapuram. In this study, the adoption of the technology coincides with uncertainty in Bantala and outright delegitimation in Nandanapuram. This uncertainty and delegitimation arise from the ambiguity related to the technology's contribution to animal deaths and cultivator indebtedness, as well as concerns for its sustainability.

The analysis highlights that the complexity of the issue of farmer suicides and risk in Telangana requires a more holistic view than the approval of further Bt crops, aimed at securing productivity in the absence of a concern with social justice, would allow. Swaminathan (2011: 25) also argues that the risk of cultivators cannot be approached only from an economic angle but must also involve considerations of ecology, equity and ethics.

As part of this, a legitimate approach to farmer suicides cannot assume that Bt technology is, or can be, a sufficient or even necessary solution to the agrarian crisis given the current unequal allocation of resources, distribution of risk and embedded power structure of the rural context. A far more responsible method would be to consult with cultivators and the landless themselves who, as the current research highlights, possess a wealth of knowledge concerning their own attempts to negotiate the risks of their context. This would involve the state government in taking a far more involved approach to the villages than the villagers asserted had occurred to date.

Beck (1992: 227) argues that the issue of risk means that 'business is not responsible for something it causes, and politics is responsible for something over which it has no control.' This book suggests that it is the unequal access to resources to negotiate risk, and the unjustified exercise of power at local, national and global levels, which represent the real in risk society. This unequal access to resources continues to ensure that risk exposure is differentiated in line with power relations rather than permitting a more collective response to emerge.

The difficulties in adopting a more collective approach to risk arise in no small measure from the in-built differentiation associated with the relations of production in the neoliberal economic system itself. This was evident during the current research where grain was allowed to rot in *godowns* (warehouses) in a country in which millions remain under-nourished. The government's response to the Supreme Court that the grain could not be distributed for free as this would lower grain prices and disincentivise cultivators no doubt made economic sense; its legitimacy was highly questionable, however, from a social justice perspective given the right to justification of the hungry.

The situation suggested flaws in the economic system itself indicating its unsuitability for delivering social justice in response to risk. The fact that GM crops are being promoted from within this global economic system as essential to increasing the productivity needed to feed the world's poor (Entine 2006) is a cause for concern. The surplus grains in India suggest that the productivity problem which GM crops proponents claim the technology addresses is not the fundamental issue with which we, as a collective of humanity, are confronted.

This book argues that until humanity has overcome the issue of the differentiated exposure to risk arising from inequality in the access to key resources to negotiate risk, as well as the exercise of power which supports this, increased agricultural productivity will simply reinforce existing power relations and exacerbate environmental risk. Such a strategy will continue to permit a situation where increasingly desperate cultivators are driven to suicide as part of a bid to create wealth within highly unequal, high-risk agrarian contexts. The suicides, and the unjustified approach to risk negotiation which they suggest, will therefore remain the subject of an ongoing legitimation struggle.

As highlighted, Bt cotton is immersed within this wider struggle to problematise power relations in local, national and, increasingly, global contexts. The conflict of which Bt technology has, in many ways, become symbolic seeks to assert the right to justification of the vulnerable on the basis of the Rawlsian Difference Principle that a just social order should not 'secure attractive prospects for the wealthy unless to do so is to the advantage of those less fortunate' (Rawls [1971], 1999: 65).

The occurrence of farmer suicides suggests that the Rawlsian basic requirement of justice is not being met. As such, the suicides represent a significant threat to the legitimacy of existing structures of power, as well as the allocation of resources which defines them. The issue of animal deaths in Warangal highlights that these power relations are also of particular concern with regard to the crucial attempts to construct knowledge in the definition of risk as a local, national and global concern. The animal deaths, too, can be seen as part of the attempt to place rural India in a 'shadow space' (Vasavi 2012) within the wider success story of India's (often urbanised) narrative of economic growth. This book argues that this shadow space is not simply significant for Indian people or for the Indian government; it is of concern globally.

The broader legitimation struggle in which Bt cotton is immersed requires the development of an enlarged imaginary capable of challenging barriers which seek to atomise not only individuals but also states. As part of this, humanity is called upon to make the global connections necessary to combat risk as a practical, moral and epistemic concern, even as the state itself is challenged to protect those within its borders as a concern for its legitimacy.

The demands for a separate state in Telangana as the means to securing a more just allocation of resources for those within the region highlight the 'boundary problem' of the state identified by Fraser (2008: 22). Here, states are charged with protecting their own citizens through preserving their access to resources while simultaneously being called upon to co-operate, negotiate and share resources as part of an emergent global process of legitimation in response to risk.

These competing claims on states are likely to become more significant as exposure to risk intensifies worldwide, rendering the issue of securing state legitimacy increasingly problematic. Attempts by states to mitigate exposure to risk without a significant redistribution of resources will lead to enhanced efforts to secure legitimacy through scientific advances. As the issue of Bt technology highlights, however, much relies upon the way in which the power relations associated with such knowledge construction itself are negotiated and justified. This will now be explored as the final section of this book.

The Construction of Knowledge and the Global Struggle for Legitimation

Renn (2008: 335) argues that issues concerning risk represent 'differences between visions of the future, basic values and convictions, and the degree of confidence in the human ability to control and direct its own technological destiny.' This relates critically to the way in which technological innovations are defined and applied, most notably in terms of the role of scientific knowledge in shaping humanity's future as a collective, and the values which should normatively inform such knowledge construction. Within this, science is recognised as crucial to identifying and resolving risk, even as it is regarded as largely responsible for creating it. Thus, as Beck (1992: 163) argues, 'science...provides the prerequisites for "overcoming" the threats for which it is responsible itself.'

Gruère and Sengupta (2011: 334) assert 'the critical need to distinguish the effect of Bt cotton as a technology from the context in which it was introduced.' This view fails to acknowledge, however, the way in which the application of scientific knowledge is presented as the means to addressing the risks of the particular contexts which these authors are urging should be separated from the technology's assessment. The view of Gruère and Sengupta also appears to suggest that the risks arising from the application of a technology within particular socio-economic contexts should be disregarded by those involved in the legitimation struggle entailed in deciding whether or not to adopt the technology. This is where Gruère and Sengupta's (2011: 316) emphasis of the *overall* effectiveness of Bt technology, rather than on its differentiated impact in particular contexts, becomes problematic.

Within Gruère and Sengupta's assertion that the technology should be distinguished from the context, there is the significant potential for the risk discourse of cultivators themselves to be marginalised as part of the emphasis on decontextualised scientific knowledge. This marginalisation of risk relates to the expert-lay divide explored by Lidskog (2008: 69), and the 'uncomfortable knowledge' (Rayner 2012: 107) which analysis of the impact of technological adoption in particular contexts can uncover.

As Lidskog (2008: 73) notes, this potential for the marginalisation of perspectives within attempts to define scientific knowledge in relation to risk has contributed to a situation where the 'capacity of science to deliver trustworthy knowledge is currently contested.' In this sense, the results of the proposed GM field trials in Telangana are likely to be delegitimated by opponents because they will be perceived as being covertly arranged and driven by a state-multinational corporation power nexus which seems to care little for the real needs of cultivators. This negative response will have been exacerbated by the fact that the trials have bypassed any attempt to secure their legitimation through the reason-giving which their justification would have entailed.

Aronowitz (1988: 32) notes that 'the production of knowledge is a social process.' The idea that, as such, it is mediated by power relations is asserted by Foucault (1977: 27) who argues that 'power and knowledge directly imply one another.' The realisation not merely that knowledge is power but that knowledge is itself defined through power has led Szerszynski (1996: 117) to assert the 'self-consciousness of a humanity becoming aware of the radical contingency of all knowledge claims.' This has meant that knowledge construction has itself become subject to legitimation challenges.

Given the 'epistemic gap' (Desmond 2014: 13) which risk represents and the significant implications of negotiating this for humanity's ongoing survival, knowledge construction is particularly crucial in relation to risk. The political dimension of knowledge construction is recognised by Visvanathan and Parmar (2002: 2724) who note that, for 'Indian democracy to be sustainable [it] still needs to understand risk.' Within this, the authors highlight that the controversy surrounding biotechnology represents 'a great moral debate' (ibid.).

Through this debate and the legitimation struggle involved in the attempt to construct knowledge with regard to risk, Indian society is being constituted, as are societies worldwide. Agreement on where research should be focussed and how knowledge should be used involves societies in reaching agreement on the type of values which should inform their future development. The recognition that these decisions concerning knowledge are themselves mediated through power relations relates to the 'politics of knowledge' identified by Beck (1992: 51). The conflict regarding knowledge construction itself could be seen in the response to the Indo-US Knowledge Initiative which Raina (2006: 1624) argues has resulted in 'enslaving the formal scientific components of the Indian agricultural knowledge system to the demands of a globally powerful player [the United States government].'

The legitimation struggle with regard to scientific knowledge relates to the capacity to critically distinguish between a legitimate and illegitimate exercise of power in both its construction and application. This research has highlighted that knowledge construction with regard to risk is deemed legitimate when the right to justification of the vulnerable and potentially marginalised is recognised and represented within its construction as a concern for epistemic injustice (Fricker 2007). This requires an exercise of power which demands that the perspectives on risk of scientists and lay people alike are treated with respect in order to aid mutual understanding given the recognised 'non-knowing' (Beck 2009: 115) which risk represents. Lidskog (2008: 73), too, notes that 'there seems to be a near consensus that greater involvement by non-experts is needed in risk assessment.'

The situation with regard to the animal deaths in Bantala highlights the problems associated with the involvement of Lidskog's 'non-experts' in practice. Here, the wider power relations involved in knowledge construction have served to reinforce the marginalisation of cultivators and to deny them in their *'capacity as…knower[s]'* (Fricker 2007: 20) [italics in the original]. This misrecognition of cultivators and the failure to address their right to justification represent a significant concern given the efforts to define the risk of Bt technology worldwide.

Clark (2007: 210) argues there is an 'emerging reality of world society. We feel its presence through the alternative normative principles that it enshrines, however embryonic and unsettled these might remain.' Clark (ibid.) asserts that '[n]ew norms will emerge from this process of negotiation as power-holders are obliged to accommodate some of the demands as a concern for their own legitimacy.'

Within world risk society, this book argues that there is an emergent global process of legitimation. The issue of GM crops can be seen as symbolic of this given that it has drawn worldwide attention to the issue of power relations in scientific knowledge construction. The concern surrounding GM crops has been reinforced by the recognition of the global nature of the risk which the technology potentially represents.

Strydom (2002: 43) argues that the 'dynamics involving both power exertion and conflict…constitute a process of construction of both national and global society…which is more fundamental than regular political competition and economic conflict.' This book claims that such conflict represents a struggle for legitimation and asserts that the exploration undertaken here of how this has materialised in the local contexts of three Indian villages can be extrapolated onto a global level.

Within this emergent global legitimation process, the right to justification of the vulnerable with regard to risk expoure and knowledge construction is asserted by international NGOs (INGOs). The increasing involvement of INGOs in deliberations in global institutions, such as the United Nations, is noted by Kuper (2004: 175) who highlights that '1,600 NGOs have consultative status [in the United Nations],' and '20 NGOs meet with the UN Security Council' (ibid.: 176). This NGO engagement is contributing to a global legitimation struggle, not only in relation to GM crops but also with regard to the negotiation of risk more generally. The efforts of such INGOs to pressurise state governments to take action on global issues heighten the sense of interconnectedness and interdependence of risk society and contribute to the constitution of global society itself.

There are undoubtedly issues of representation with regard to the engagement of INGOs in international deliberations. There are also concerns with the way in which such INGO involvement may serve to legitimate the global power structure in the same way as the NGO engagement endorsed by Pradnesh in Orgampalle served to consolidate the legitimation of his power in the village; nonetheless, at a global level, the involvement of NGOs in deliberations contributes to efforts to assert recognition for the right to justification of the vulnerable in local contexts. Their engagement also serves to ensure the inclusion of issues of global justice within national legitimation struggles. Here, problems related to the exercise of power by individual states become problematised at a global level in ways which can lead to their being raised as issues by citizens as part of the process of legitimation within those states.

The global legitimation process concerning the issue of GM crops has contributed not only to demands for 'epistemological decentralisation' (Pieterse 2001: 89); it has also created the impetus for an epistemological cosmopolitanisation which has seen an enhanced sharing and challenging of knowledge construction worldwide. This was seen in the case of the two NGOs in the current study who exchanged local information on Bt cotton cultivation with NGO networks globally.

Beck (2009: 114) argues that risk society cannot 'be overcome by more and better science.' This refers to the need to guard against the reductionist scientism which, as Roderick (1986: 50) notes, means that 'we no longer understand science as one form of possible knowledge but rather identify knowledge with science.' Thus, our need to ensure that the sources from which we gather information in knowledge construction on risk are as broad and diverse as possible represents the key to our negotiation of risk.

The legitimation struggle concerning Bt technology illustrates how the potential for scientific reductionism is countered from within the paradigm of science itself by the emergence of 'adversarial science' (Renn 2008: 217). This represents an attempt to re-assert the importance of context within a field which has for so long emphasised universal, value-free, 'pure' knowledge. Challenges to the idea of 'pure' knowledge have seen an increased demand for 'postnormal science' (Lidskog 2008: 76) with a greater emphasis on 'specialisation in the context' (Beck 1992: 178). Such a concern forms the basis of the protests and legitimation challenges related to risk worldwide. As this book has highlighted, however, care also needs to be taken with regard to the impact of power relations on the construction of localised knowledge and measures adopted to secure that the perspectives of the marginalised are recognised within this.

Beck (2009: 210–211) notes the need for ongoing critical sociological cross-cultural analyses on risk in particular contexts which avoid tendencies towards 'methodological nationalism.' In this way, humanity can start to develop a knowledge base which is genuinely informed by legitimation struggles worldwide, and which serves to inform, direct and guide knowledge construction on risk, and the uses to which scientific and technological innovations are applied in particular contexts globally.

The interconnectedness and interdependence of risk society, and the focus on legitimation which it entails, hold significant potential for pushing the boundaries of epistemology in response to risk. My own experience in producing this book is testament to the incredible opportunity for learning which risk society presents, and which travel and media technologies support. This learning opportunity relates not just to understandings of risk as a theoretical concern but also as a moral one.

These efforts to expand the boundaries of learning bring awareness of the lived reality of risk, and the legitimation struggles which risk gives rise to within particular contexts, even as they heighten the consciousness of the global conflict within which such local legitimation struggles are embedded. The enhanced understanding of local contexts which are not one's own contributes to an awareness of interconnectedness and lends itself to more nuanced deliberations on how risk can be resolved at local, national and global levels. As noted, the expansion of boundaries within knowledge construction relating to risk contributes to an emergent global legitimation struggle which increasingly challenges attempts by states to mitigate their own risk exposure at the expense of others.

Beck (2009: 211) argues that, in a world at risk, there is a need for greater awareness of 'the manifold, real, self-critical voices of the developing world risk society.' This book has sought to contribute to this through foregrounding the views of Indian cultivators themselves. It could be argued that the attempt at knowledge construction in relation to risk which this book represents is itself flawed as a result of the way in which it is itself mediated through the power relations involved in the research process which were noted in the methodology.

In response, I would argue that I have been fortunate enough to have had exposure to positions on both sides of the debate, as a Syngenta employee in the UK and Switzerland, as a student of Sociology in Ireland and as a researcher interacting with cultivators, NGOs, politicians, regulators, academics and industry representatives in India. The research in India upon which this book is based was funded by the Irish Research Council, a body which does not itself adopt a position within the debate, nor did it try to influence the findings in any way. I have tried to undertake the research and write this book with a critical awareness of the power relations involved at all stages and with every possible effort to limit the impact of these upon the genuine attempt to construct knowledge which this work entails.

This book signifies my best endeavour to represent these diverse positions as legitimately as I can, given the meaning that I have tried to make of this theme from these interactions and experiences. Indeed, this work asserts the incorporation of these diverse perspectives as the basis for its own justification. It is my hope that this work will suggest avenues for ongoing research into the crucially important areas which the book has sought to highlight.

The approach of each of the three villages in this study in many ways represent future potentials in the global negotiation of risk—the concentration of power and protection of the powerful (Bantala)—fragmentation, contradiction and indecision (Nandanapuram) or a more collective response which coincides with the authoritarian tendencies of a rules-driven approach (Orgampalle). It is likely that the future negotiation of risk as a global concern will encompass aspects of all three approaches and that these will be challenged and re-worked through legitimation struggles at local, national and global levels.

It is also possible, of course, that a new approach will emerge which transcends those explored in the villages. This would entail a negotiation of risk which took the basis of its legitimation from the consciousness of the 'common human vulnerability that emerges with life itself' (Butler 2004: 31). Such a consciousness would require the development of an 'enlarged mentality' (Arendt [1954], 2006: 237) in which individual attempts to negotiate risk were informed by concerns for the security of the collective of humanity. This approach would involve a fundamental recognition that we live in a world where exposure to risk is pervasive and cannot be effectively combated either through individuals or states working in isolation. The emergence of such a consciousness is, this book argues, contingent upon the outcome of legitimation struggles being undertaken throughout the world.

The unfolding of these struggles is strongly dependent upon the future materialisation of risk and the way in which power-holders respond to this as a concern for the legitimacy of their positions. This will draw increasing attention to Banerjee's (2007: 170) view that '[t]he questions that need to be asked are…how do we understand ourselves and our world and how should we negotiate our relationships with ourselves?' These are questions which will continue to form the core concern of legitimation struggles worldwide as humanity seeks to define a legitimate exercise of power and construction of knowledge capable of securing its own survival within a world at risk.

Bibliography

Arendt, H. ([1954], 2006). *Between past and future.* London: Penguin Books.

Aronowitz, S. (1988). *Science as power: Discourse and ideology in modern society.* Minneapolis: University of Minnesota Press.

Banerjee, S. B. (2007). Who sustains whose development? Sustainable development and the reinvention of nature. *Organization Studies, 24*(1), 143–180.

Beck, U. (1992). *Risk society: Towards a new modernity.* London: Sage.

Beck, U. (1995). *Ecological politics in an age of risk.* Cambridge: Polity Press.

Beck, U. (2009). *World at risk.* Cambridge: Polity Press.

Butler, J. (2004). *Precarious life: The powers of mourning and violence.* London: Verso.

Clark, I. (2007). Legitimacy in international or world society. In A. Hurrelmann, S. Schneider, & J. Steffek (Eds.), *Legitimacy in an age of global politics* (pp. 193–210). New York: Palgrave Macmillan.

Desmond, E. (2014). *The legitimation of risk and democracy: A case study of Bt cotton in Andhra Pradesh, India.* Cork: University College Cork. Available at: https://cora.ucc.ie/handle/10468/1688/

Entine, J. (Ed.). (2006). *Let them eat precaution: How politics is undermining the genetic revolution in agriculture.* Washington, DC: American Enterprise Institute.

Foucault, M. (1977). *Discipline and punish.* London: Penguin.

Fraser, N. (2008). *Scales of justice: Reimagining political space in a globalizing world.* Cambridge: Polity Press.

Fricker, M. (2007). *Epistemic injustice: Power & the ethics of knowing.* Oxford: Oxford University Press.

Gaurav, S., & Mishra, S. (2012). *To Bt or not to Bt? Risk and uncertainty considerations in technology assessment.* Mumbai: Indira Gandhi Institute of Development Research.

Gruère, G., & Sengupta, D. (2011). Bt cotton and farmer suicides in India: An evidence-based assessment. *Journal of Development Studies, 47*(2), 316–337.

Herring, R. J. (2008). Whose numbers count? Probing discrepant evidence on transgenic cotton in the Warangal district of India. *International Journal of Multiple Research Approaches, 2*(2), 145–159.

Jakimow, T. (2014). Gambling on livelihoods: Desire, hope and fear in agrarian Telangana. *Asian Journal of Social Science, 42*, 409–434.

Jodhka, S. (2005). *Beyond 'crisis': Rethinking contemporary Punjab agriculture, Working paper* (Vol. 4). Hyderabad: CESS.

Kuper, A. (2004). *Democracy beyond borders: Justice and representation in global institutions.* Oxford: Oxford University Press.

Lidskog, R. (2008). Scientised citizens and democratised science. Re-assessing the expert-lay divide. *Journal of Risk Research, 11*(1–2), 69–86.

Pieterse, J. N. (2001). *Development theory: Deconstructions/reconstructions.* New Delhi: Vistaar.

Raina, R. S. (2006). Indo-US knowledge initiative: Need for public debate. *Economic and Political Weekly, 41*(17), 1622–1624.

Rawls, J. ([1971], 1999). *A theory of justice: Revised edition.* Cambridge: Harvard University Press.

Rayner, S. (2012). Uncomfortable knowledge: The social construction of ignorance in society and environmental policy discourses. *Economy and Society, 41*(1), 107–125.

Renn, O. (2008). *Risk governance: Coping with uncertainty in a complex world.* London: Earthscan.

Roderick, R. (1986). *Habermas and the foundations of critical theory.* London: Macmillan.

Sridhar, V. (2006). Why do farmers commit suicide? The case of Andhra Pradesh. *Economic and Political Weekly, 41*(16), 1559–1565.

Stone, G. D. (2011). Field *versus* farm in Warangal: Bt cotton, higher yields, and larger questions. *World Development, 39*(3), 387–398.

Strydom, P. (2002). *Risk, environment and society.* Buckingham: Open University Press.

Swaminathan, M. S. (2011). *In search of biohappiness: Biodiversity and food, health and livelihood security.* London: World Scientific.

Szerszynski, B. (1996). On knowing what to do: Environmentalism and the modern problematic. In S. Lash, B. Szerszynksi, & B. Wynne (Eds.), *Risk, environment & modernity: Towards a new ecology.* London: Sage.

Vasavi, A. R. (2010). Contextualising the agrarian suicides. In R. S. Deshpande & S. Arora (Eds.), *Agrarian crisis and farmer suicides* (pp. 70–85). New Delhi: Sage.

Vasavi, A. R. (2012). *Shadow space: Suicides and the predicament of rural India.* New Delhi: Three Essays Collective.

Visvanathan, S., & Parmar, C. (2002). A biotechnology story: Notes from India. *Economic and Political Weekly, 37*(27), 2714–2724.

Walker, T. S., & Ryan, J. G. (1990). *Village and household economies in India's semi-arid tropics.* London: The Johns Hopkins University Press.

Weber, M. ([1968], 1978). *Economy and society: An outline of interpretative sociology* (Vol. 1). London: University of California Press.

Appendix 1: Consent Form—Village *Sarpanch*

Research Approval

(Verbally translated into Telugu)

I _____ (sarpanch of …) confirm that I agree that the village of **…** can be used as a research location for the Ph.D. project of Elaine Desmond from University College Cork, Ireland.

I understand this project will look at Bt cotton and will involve interviews with selected villagers in order to discover the different impacts of the crop for various households within the village.

It has been explained that the identities of the village and the participants in the research will be protected.

It has also been explained that information gathered will be used for the purposes of the Ph.D. research, academic papers and creative projects and that the anonymity of the participants will be respected throughout.

SIGNED: _____   DATE: _____
RESEARCHER: _____   DATE: _____
WITNESS: _____   DATE: _____

© The Author(s) 2018
E.L. Desmond, *Legitimation in a World at Risk*,
https://doi.org/10.1007/978-981-10-6065-6

Appendix 2: Consent Form—Participants

Consent Form

(Verbally translated into Telugu at the beginning of the research)

I _____ agree to participate in Elaine Desmond's research study on GM crops.

The purpose and nature of the study have been explained to me in writing.

I am participating voluntarily.

I give permission for my interviews with Elaine Desmond to be tape-recorded.

I understand that I can withdraw permission to use the data within 2 weeks of the start of the research, in which case any material collected will be destroyed.

I understand that anonymity will be ensured in the write-up by disguising my identity.

© The Author(s) 2018
E.L. Desmond, *Legitimation in a World at Risk*,
https://doi.org/10.1007/978-981-10-6065-6

I understand that disguised extracts from my interview may be quoted in the thesis and any subsequent publications if I give permission below:

(Please tick one box)

I agree to quotation/publication of extracts from my interview.   □

I do not agree to quotation/publication of extracts from my interview. □

Signed _____   Date _____

Appendix 3: Confidentiality Agreement with Translator

Confidentiality Agreement

I confirm that the identities of research participants to whom I am introduced as part of my work as translator for Elaine Desmond will be respected.

I confirm that the identity of the village(s) where this research is based will be kept confidential.

I confirm that information which is disclosed to me as part of this work will remain confidential and that the anonymity of research participants will be respected.

SIGNED: _____

DATE: _____

© The Author(s) 2018
E.L. Desmond, *Legitimation in a World at Risk*,
https://doi.org/10.1007/978-981-10-6065-6

Appendix 4: Village Composition— Bantala, 2010/2011

Caste	Traditional occupation	Number of households	Land-holding in acres per caste[a]
Forward Castes			
Reddy	Agriculture	8	70
Vaishya	Merchants, land-owners, moneylenders	4	40
Backward Castes			
Kuruma[b]	Shepherds	90	1100
Gowda	Cattle breeders/toddy tappers	80	250
Are	Agriculture	10	100
Vadrangi	Carpenters	10	90
Chakali	Washermen	20	80
Kapu/Munnuru Kapu	Land-owners	10	70
Kumari	Potter	5	25
Kamari	Blacksmith	8	20
Mangali	Barber	5	20
Padmashali	Weavers	8	15

(continued)

© The Author(s) 2018
E.L. Desmond, *Legitimation in a World at Risk*,
https://doi.org/10.1007/978-981-10-6065-6

(continued)

Caste	Traditional occupation	Number of households	Land-holding in acres per caste[a]
Scheduled Castes			
Madiga[c]	Leather workers	150	300 (of which only 150 acres cultivated)[d]
Mala	Weaving, watchmen, agricultural labourers	20	50
Total		428	2120

[a]Ranked according to land-holding within caste categories
[b]Dominant caste and Kuruma *sarpanch*
[c]Majority population
[d]This includes uncleared land from land reform which has been allocated to the Madigas but which is unfit for cultivation

Appendix 5: Village Composition— Nandanapuram, 2010/2011

Caste	Traditional occupation	Number of households	Land-holding in acres per caste[a]
Forward Castes			
Reddy	Agriculture[b]	20	100
Vaishya	Merchants, land-owners, moneylenders	5	15
Brahmins	Priests, teachers, administrators	2	14
Backward Castes			
Gowda[c]	Cattle breeders/toddy tappers	200	496
Boya	Hunters	250	300
Kumari	Potter	40	200
Yadava	Cattle herding	22	50
Vadla	Carpenters	20	20
Chakali (dhobi)	Washermen	10	5
Darji	Tailors	2	2

(continued)

© The Author(s) 2018
E.L. Desmond, *Legitimation in a World at Risk*,
https://doi.org/10.1007/978-981-10-6065-6

(continued)

Caste	Traditional occupation	Number of households	Land-holding in acres per caste[a]
Scheduled Castes			
Madiga	Leather workers	400[d]	700
Total		971	1902

[a]Ranked according to land-holding within caste categories

[b]Although the land-holding of Reddys in Nandanapuram is small relative to Gowdas, Boyas, Kumaris and Madigas in the village, there are large land-holding individuals within the Reddy *jati*. They are also identified by villagers as being powerful, though less so than previously given the empowerment of the SC Madigas and BC Gowdas. This relates to the wider power with which the *jati* is associated in Telangana and highlights the way in which power relations in the villages intersect with those in the wider state

[c]The *sarpanch* is a female BC Gowda

[d]Scheduled Caste Madigas are numerically in the majority in Nandanapuram but are not identified as dominant given their low caste ranking and employment as agricultural labourers. Instead, power is negotiated between Backward Caste Gowdas, Boyas and Reddys, as well as Scheduled Caste Madigas

Appendix 6: Village Composition— Orgampalle, 2010/2011

Caste	Traditional occupation	Number of households	Land-holding in acres per caste
Backward Castes			
Mudhiraj[a]	Agriculture	48	236
Yadava	Cattle herding	2	25
Chakali	Washermen	2	25
Total		52	286

The *sarpanch* (a male BC Mannuru Kappu) is located in a neighbouring village
[a]Dominant caste

© The Author(s) 2018
E.L. Desmond, *Legitimation in a World at Risk*,
https://doi.org/10.1007/978-981-10-6065-6

Appendix 7: 'Rules' of Orgampalle Village

(Translation from Telugu)

1. We farmers follow organic cultivation methods.
2. We prepare our own seeds.
3. We grow mixed crops.
4. Every household grows trees.
5. We grow vegetables in the backyard of every house.
6. We will develop the dairy industry.
7. In every family, the farmer will be a member of the Self-Help Group.[1]
8. Every farmer will maintain a register giving details of how their crops are cultivated.
9. Farmers will abide by the conditions agreed upon by the village society and Crops Jangaon.
10. We ban GM crops.
11. We will make other farmers aware of organic methods.
12. We will dig no more borewells.
13. Every farmer who has sufficient water will share it with others.

Sky water (rain) is sacred, river water is more sacred, and ground water is the most sacred. We will conserve rain water and increase ground water.

© The Author(s) 2018
E.L. Desmond, *Legitimation in a World at Risk*,
https://doi.org/10.1007/978-981-10-6065-6

Notes

1. Organised by Crops Jangaon.

Appendix 8: Caste Wards and Locations of Participants—Bantala, 2010/2011

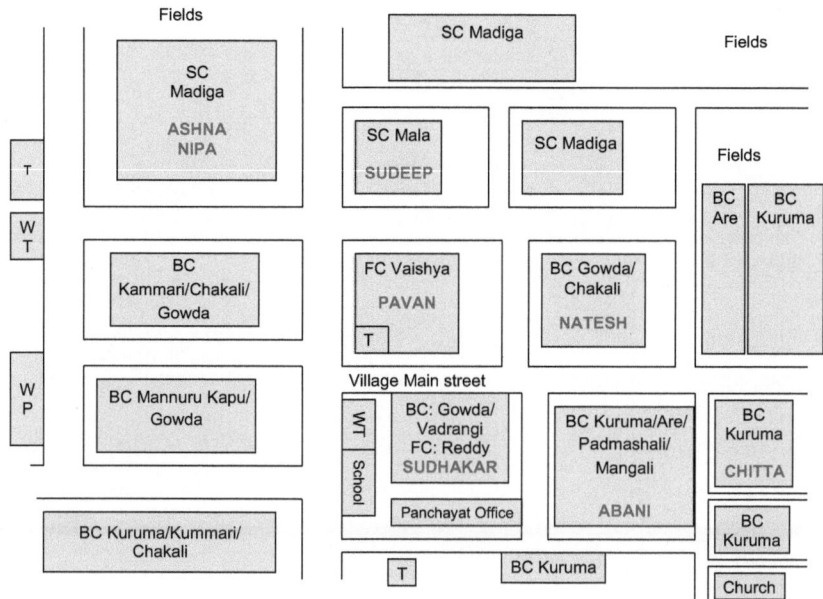

BANTALA
2010/2011

Fields	SC Madiga	Fields

SC Madiga
ASHNA NIPA

SC Mala
SUDEEP

SC Madiga

Fields

T
W T

BC Are | BC Kuruma

BC Kammari/Chakali/Gowda

FC Vaishya
PAVAN
T

BC Gowda/Chakali
NATESH

Fields

W P

BC Mannuru Kapu/Gowda

Village Main street

WT

BC: Gowda/Vadrangi
FC: Reddy
SUDHAKAR

School

Panchayat Office

BC Kuruma/Are/Padmashali/Mangali

ABANI

BC Kuruma
CHITTA

BC Kuruma

BC Kuruma/Kummari/Chakali

T

BC Kuruma

Church

FC: Forward (upper) Caste;
BC: Backward Caste;
SC: Scheduled Caste;
T – Temple; WP – Water Purifier; WT – Water Tank; Church for small Christian minority.

© The Author(s) 2018

E.L. Desmond, *Legitimation in a World at Risk*,
https://doi.org/10.1007/978-981-10-6065-6

Appendix 9: Caste Wards and Locations of Participants—Nandanapuram, 2010/2011

NANDANAPURAM
2010/2011

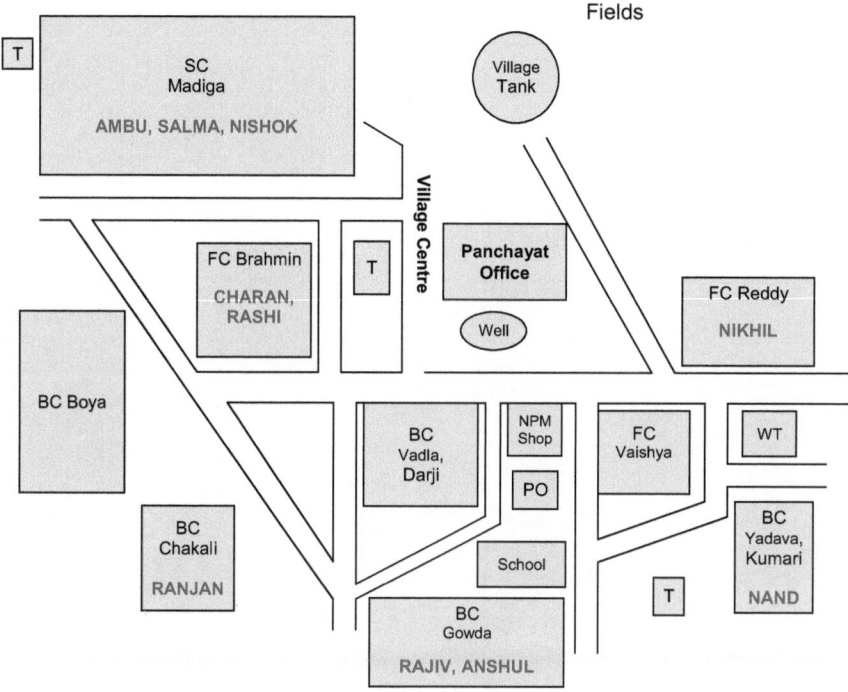

FC: Forward (upper) Caste;
BC: Backward Caste;
SC: Scheduled Caste;
T – Temple; WT – Water Tank; PO – Post Office

© The Author(s) 2018
E.L. Desmond, *Legitimation in a World at Risk*,
https://doi.org/10.1007/978-981-10-6065-6

Appendix 10: Caste Wards and Locations of Participants—Orgampalle, 2010/2011

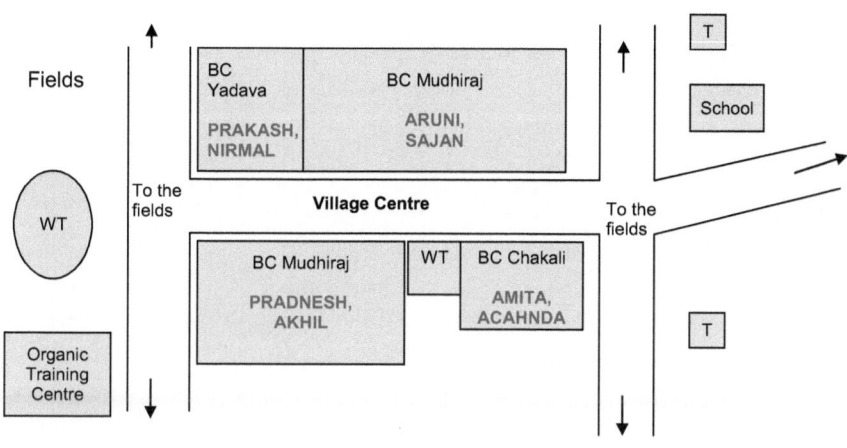

ORGAMPALLE
2010/2011

BC: Backward Caste;
T – Temple; WT – Water Tank

© The Author(s) 2018
E.L. Desmond, *Legitimation in a World at Risk*,
https://doi.org/10.1007/978-981-10-6065-6

Appendix 11: Risk Profile of Bantala Participants

© The Author(s) 2018
E.L. Desmond, *Legitimation in a World at Risk*,
https://doi.org/10.1007/978-981-10-6065-6

Name[a]	Caste	Gender	Total land-holding	Land allocated to cotton	Owns borewell?	Tractor/oxen ownership	Off-farm income
Pavan	Forward Caste (FC Vaishya)	Male	20 acres (17 leased)	20 acres	Yes	Oxen	Village shop and rice milling machine
Sudhakar	Forward Caste (FC Reddy)	Male	16 acres (4 leased)	5 acres	Yes	Tractor	No
Chitta	Backward Caste (BC dominant)	Male	10 acres (5 leased)	7 acres	Yes	Oxen	Director of Co-operative Bank
Natesh	Backward Caste (BC Gowda)	Male	6.5 acres (4 leased)	4 acres	No	Yes[b]	None
Ashna	Scheduled Caste (SC Madiga)	Female	3 acres	2 acres	No	No	Husband in construction
Sudeep	Scheduled Caste (SC Mala)	Male	1 acre	1 acre	No	No	Auto-rickshaw driver
Abani	Backward Caste (BC Padmashali)[c]	Female	Landless	–	–	–	Agricultural labourer
Nipa	Scheduled Caste (SC Madiga)	Female	Landless	–	–	–	Agricultural labourer

[a]Ranked in order of relative power in the village

[b]Sold during 2010/2011 season to pay agricultural labour

[c]The traditional occupation of the Padmashali caste *jati* is weaving. Despite being classified as an 'Other Backward Caste' due to their economic deprivation within the Backward Caste category, the *jati* is noteworthy ritually, with males of the caste wearing the sacred thread or *suta* traditionally reserved for Brahmins. This explains the location of Abani's residence close to the centre of the village in Appendix 8

Appendix 12: Risk Profile of Nandanapuram Participants

© The Author(s) 2018
E.L. Desmond, *Legitimation in a World at Risk*,
https://doi.org/10.1007/978-981-10-6065-6

Name[a]	Caste	Gender	Total land-holding	Land allocated to cotton	Owns borewell?	Tractor/oxen ownership	Off-farm income
Nikhil	Forward Caste (Reddy)	Male	10.5 acres	10.5 acres	Yes	Oxen and tractor	None
Charan	Forward Caste (Brahmin)	Male	12 acres	8 acres	No	Oxen	None
Rajiv	Backward Caste (Gowda)	Male	4 acres (2 acres leased)	4 acres	Yes	No	None
Nand	**Backward Caste (Yadava)**	Male	**6 acres**[b]	**3 acres**	No	No	NGO employee
Anshul	Backward Caste (Gowda)	Male	5 acres	2 acres	No	No	None
Nishok	**Scheduled Caste (Madiga)**	Male	**3 acres**[b]	**2 acres**	No	No	None
Rashi	Forward Caste (Brahmin)	Female	4 acres	2 acres	No	No	Sons' incomes as priests
Ambu	Scheduled Caste (Madiga)	Female	4 acres	3 acres	Yes (shares with brothers-in-law)	Oxen	Widow's pension
Ranjan	Backward Caste (Chakali)	Male	Landless	–	–	–	Laundry work and pension (now held back by *sarpanch*)
Salma	Scheduled Caste (Madiga)	Female	Landless	–	–	–	Agricultural labourer

[a]Ranked in order of relative power in the village
[b]NPM cultivators

Appendix 13: Risk Profile of Orgampalle Participants

© The Author(s) 2018
E.L. Desmond, *Legitimation in a World at Risk*,
https://doi.org/10.1007/978-981-10-6065-6

Name[a]	Caste	Gender	Total land-holding	Land allocated to cotton	Owns borewell?	Tractor/oxen ownership	Off-farm income
Pradnesh	Backward Caste (dominant)	Male	20 acres	4 acres	No	Tractor and oxen	None
Akhil	Backward Caste (dominant)	Male	5 acres	1.5 acres	No	Oxen	None
Nirmal	Backward Caste (Yadava)	Male	5 acres	3 acres	No	Oxen	None
Achanda	Backward Caste (Chakali)	Male	4 acres (all leased)	3 acres	Yes[b]	Oxen	Remittances from son in nearby town
Prakash	Backward Caste (Yadava)	Male	1.5 acres	1.5 acres	No	Oxen	NGO employee
Aruni	Backward Caste (dominant)	Female	4 acres	1.5 acres	No	Oxen	Husband a herder
Amita	Backward Caste (Chakali)	Female	4 acres	1 acre	No	No	Widow's pension
Ranjan	Backward Caste	Male	Landless	–	–		Pension

[a]Ranked in order of relative power in the village
[b]Not used for cotton. Drilled prior to the ban on borewells in village

Appendix 14: Debt Levels per Participant—Bantala, 2010/2011

© The Author(s) 2018
E.L. Desmond, *Legitimation in a World at Risk*,
https://doi.org/10.1007/978-981-10-6065-6

	Caste	Gender	Total land-holding	Debt level	Main reasons given for borrowing	Sources of debt
Average debt level for Bt cotton cultivators				227,500ᵃ		
Pavan (FC)	Forward Caste (FC Vaishya)	Male	20 acres (17 leased)	400,000	Cultivation costs (seeds, inputs, labour), dowry, education of children	Banks; people in other villages
Sudhakar (FC)	Forward Caste (FC Reddy)	Male	16 acres (4 leased)	700,000	Buying land; borewell; tractor; cattle shed	Bank; people in own villages
Chitta (BC)	Backward Caste (BC dominant)	Male	10 acres (5 leased)	150,000	Borewell; cultivation costs (seeds, inputs, labour)	Banks; private money lender in another village
Natesh (BC)	Backward Caste (BC Gowda)	Male	6.5 acres (4 leased)	65,000	Dowry for sisters and daughter; cultivation costs; son's private hospital treatment	Bank; people in own village and other villages
Ashna (SC)	Scheduled Caste (SC Madiga)	Female	3 acres	50,000	Labour charges (incurred this season); daughter's treatment in private hospital	Bank; micro-finance institution
Sudeep (SC)	Scheduled Caste (SC Mala)	Male	1 acre	0		
Abani (BC)	Backward Caste (BC Padmashali)	Female	Landless	100,000	Private hospital—husband's surgery	Others in own village
Nipa (SC)	Scheduled Caste (SC Madiga)	Female	Landless	24,000	Private hospital; Andhra Pradesh rural poverty reduction project—training to work as a cobbler	Bank (land-owner as guarantor); 'landlords' in the village

ᵃDebt levels of landless participants are not included in the average debt level due to the emphasis on the contribution of Bt cotton cultivation to debt levels (N = 6)

Appendix 15: Debt Levels per Participant— Nandanapuram, 2010/2011

© The Author(s) 2018
E.L. Desmond, *Legitimation in a World at Risk*,
https://doi.org/10.1007/978-981-10-6065-6

	Caste	Gender	Total land-holding	Debt level	Main reasons given for borrowing	Sources of debt
Average debt level of village participants				**NP (Bt) Rs 191,666**[a] **NP (NPM) Rs 12,500**		
Nikhil (FC)	Forward Caste (Reddy)	Male	10.5 acres	250,000	Cultivation costs; education of children	Commission agent in Warangal; bank
Charan (FC)	Forward Caste (Brahmin)	Male	12 acres	110,000	Input and labour costs	Bank; person in the village
Rajiv (BC)	Backward Caste (Gowda)	Male	4 acres (2 acres leased)	250,000	Dowry; cultivation costs (inputs, seeds, labour)	Own village
Nand (BC) (NPM)	**Backward Caste (Yadava)**	**Male**	**6 acres**	**0**		
Anshul (BC)	Backward Caste (Gowda)	Male	5 acres	80,000	Input and labour costs	Trader in another village
Nishok (SC) (NPM)	**Scheduled Caste (Madiga)**	**Male**	**3 acres**	**25,000**	Labour costs (current season)	Own village
Rashi (FC)	Forward Caste (Brahmin)	Female	4 acres	350,000	House construction; health; dowry (of sister's daughters)	Bank; people in own village (own caste)
Ambu (SC)	Scheduled Caste (Madiga)	Female	4 acres	110,000	Dowry; education of children; hospital bills; input and labour costs	Own village; (Gowda caste) and people from other villages; sister Family. Sold buffalo to repay
Ranjan (BC)	Backward Caste (Chakali)	Male	Landless	0	Previous debt due to health costs	
Salma (SC)	Scheduled Caste (Madiga)	Female	Landless	10,000	Lease of land for paddy cultivation and crop failure	Self-Help Group

[a]Does not include debt of landless participants (*N* = 6)

Appendix 16: Debt Levels per Participant—Orgampalle, 2010/2011

© The Author(s) 2018
E.L. Desmond, *Legitimation in a World at Risk*,
https://doi.org/10.1007/978-981-10-6065-6

Caste	Gender	Total land-holding	Debt level (Rs)	Main reasons given for borrowing	Sources of debt
Average debt level of village participants			**42,375**		
Pradnesh (BC) Backward Caste (dominant)	Male	20 acres	0		People in nearby village
Akhil (BC) Backward Caste (dominant)	Male	5 acres	4000	Labour costs (2010/2011 season)	People in nearby village
Nirmal (BC) Backward Caste (Yadava)	Male	5 acres	100,000	House construction	Cross-cousin and people in other villages
Achanda (BC) Backward Caste (Chakali)	Male	4 acres (all leased)	50,000	Labour charges in the 2010/2011 season. Dowry for daughters (1 lakh for one daughter and Rs 50,000 for another)	Bank. People in own and other villages
Prakash (BC) Backward Caste (Yadava)	Male	1.5 acres	35,000	House construction	Bank and NGO society
Aruni (BC) Backward Caste (dominant)	Female	4 acres	50,000	Dowry of two daughters. Borewell (years ago). Since dried up	Bank; MFI in nearby town
Amita (BC) Backward Caste (Chakali)	Female	4 acres	100,000	Dowry of four daughters; health	People from nearby villages
Sajan (BC) Backward Caste	Male	Landless	0		

Appendix 17: Questionnaire Results—2010/2011

1. I want to grow Bt cotton.

	Bantala (%) (N = 6)	Nandanapuram (%) (N = 8)	Orgampalle (%) (N = 7)
Strongly agree	66		
Agree	17	12	
Don't know	17	14	
Disagree			
Strongly disagree		62	58
Unaware of Bt cotton		12	42

2. Growing Bt cotton has increased my household income.

	Bantala (%) (N = 6)	Nandanapuram (%) (N = 8)	Orgampalle (%) (N = 7)
Strongly agree	83		15
Agree	17	13	
Don't know			15
Disagree			
Strongly disagree		75	28
Unaware of Bt cotton		12	42

© The Author(s) 2018
E.L. Desmond, *Legitimation in a World at Risk*,
https://doi.org/10.1007/978-981-10-6065-6

3.   Growing Bt cotton has led to increased crop yields.

	Bantala (%) (N = 6)	Nandanapuram (%) (N = 8)	Orgampalle (%) (N = 7)
Strongly agree	83	25	15
Agree	17		
Don't know			15
Disagree			
Strongly disagree		63	28
Unaware of Bt cotton		12	42

4.   I have complete control of whether I grow Bt cotton or not.

	Bantala (%) (N = 6)	Nandanapuram (%) (N = 8)	Orgampalle (%) (N = 7)
Strongly agree	100	50	43
Agree			
Don't know			
Disagree		25	15
Strongly disagree		13	
Unaware of Bt cotton		12	42

5.   Bt cotton is not damaging to the environment.

	Bantala (%) (N = 6)	Nandanapuram (%) (N = 8)	Orgampalle (%) (N = 7)
Strongly agree		12	
Agree			
Don't know	83	13	15
Disagree		13	
Strongly disagree	17	50	43
Unaware of Bt cotton		12	42

6.   I believe Bt crops will make more food available to those who need it.

	Bantala (%) (N = 6)	Nandanapuram (%) (N = 8)	Orgampalle (%) (N = 7)
Strongly agree	83	13	15
Agree			15
Don't know	17	13	14
Disagree			
Strongly disagree		62	14
Unaware of Bt cotton		12	42

7.   I would have no health concerns about eating Bt brinjal (aubergine).

	Bantala (%) (N = 6)	Nandanapuram (%) (N = 8)	Orgampalle (%) (N = 7)
Strongly agree	16		
Agree			
Don't know	17	13	15
Disagree	17	25	
Strongly disagree	50	50	43
Unaware of Bt cotton		12	42

8.   Growing Bt cotton has allowed myself and my family easier access to food.

	Bantala (%) (N = 6)	Nandanapuram (%) (N = 8)	Orgampalle (%) (N = 7)
Strongly agree	33	12	15
Agree	33	13	
Don't know			15
Disagree	17	25	14
Strongly disagree	17	38	14
Unaware of Bt cotton		12	42

9.   Growing Bt cotton has made my life easier.

	Bantala (%) (N = 6)	Nandanapuram (%) (N = 8)	Orgampalle (%) (N = 7)
Strongly agree	50	13	
Agree	16	13	28
Don't know	17		15
Disagree	17	12	
Strongly disagree		50	15
Unaware of Bt cotton		12	42

10.   **Fewer pesticide sprays are required on Bt cotton.**

	Bantala (%) (N = 6)	Nandanapuram (%) (N = 8)	Orgampalle (%) (N = 7)
Strongly agree	83	38	15
Agree			15
Don't know			14
Disagree			
Strongly disagree	17	50	14
Unaware of Bt cotton		12	42

11.   **The government is interested in the experience of farmers who grow Bt cotton.**

	Bantala (%) (N = 6)	Nandanapuram (%) (N = 8)	Orgampalle (%) (N = 7)
Strongly agree	50	12	
Agree		13	15
Don't know			15
Disagree	16	13	14
Strongly disagree	34	50	14
Unaware of Bt cotton		12	42

12.   **Economic development, through science and technology, is essential for reducing poverty.**

	Bantala (%) (N = 6)	Nandanapuram (%) (N = 8)	Orgampalle (%) (N = 7)
Strongly agree	83	75	
Agree	17	13	
Don't know			15
Disagree			28
Strongly disagree			15
Unaware of Bt cotton		12	42

13. India must compete on the global economic market in order to deal with poverty.

	Bantala (%) (N = 6)	Nandanapuram (%) (N = 8)	Orgampalle (%) (N = 7)
Strongly agree	83	50	28
Agree		25	
Don't know	17	13	15
Disagree			
Strongly disagree			15
Unaware of Bt cotton		12	42

14. The democratic process has led to greater equality in terms of access to opportunities in Indian society.

	Bantala (%) (N = 6)	Nandanapuram (%) (N = 8)	Orgampalle (%) (N = 7)
Strongly agree	83	25	
Agree		25	
Don't know			16
Disagree		13	
Strongly disagree	17	25	42
Unaware of Bt cotton		12	42

Bibliography

Abromeit, H., & Stoiber, M. (2007). Criteria of democratic legitimacy. In A. Hurrelmann, S. Schneider, & J. Steffek (Eds.), *Legitimacy in an age of global politics* (pp. 35–56). New York: Palgrave Macmillan.

Adeney, K., & Wyatt, A. (2010). *Contemporary India.* New York: Palgrave Macmillan.

Agarwal, A. K. (1980). Agricultural crop insurance in India. In C. Y. Lee (Ed.), *Crop insurance for Asian countries* (pp. 99–113). Bangkok: Food and Agricultural Organisation Regional Office for Asia and the Pacific.

Agarwal, B. (1995). Gender and legal rights in agricultural land in India. *Economic and Political Weekly, XXX*(12), A-39–A-56.

Agarwal, B. (2003). Gender and land rights revisited: Exploring new prospects via the state, family and market. *Journal of Agrarian Change, 3*(1 and 2), 184–224.

Agnihotri, V. K., & Subramanian, S. V. (2002). Andhra Pradesh. In V. K. Agnihotri (Ed.), *Socio-economic profile of rural India* (Vol. I). New Delhi: Concept Publishing Company.

Alam, J. (2004). What is happening inside Indian democracy? In R. Vora & S. Palshikar (Eds.), *Indian democracy: Meanings and practices.* London: Sage Publications.

© The Author(s) 2018
E.L. Desmond, *Legitimation in a World at Risk,*
https://doi.org/10.1007/978-981-10-6065-6

Ali, F., Panikker, K. M., & Kunzru, H. N. (1955). *Report of the states reorganisation commission.* Delhi: Government of India.

Arendt, H. ([1954], 2006). *Between past and future.* London: Penguin Books.

Aronowitz, S. (1988). *Science as power: Discourse and ideology in modern society.* Minneapolis: University of Minnesota Press.

Ayyangar, S. (2003). Janmabhoomi meetings in two villages. *Economic and Political Weekly, 38*(42), 4426–4429.

Bandiera, O., & Rasul, I. (2006). Social networks and technology adoption in Northern Mozambique. *The Economic Journal, 116*(514), 869–902.

Bandyopadhyay, D. (2001). Andhra Pradesh: Looking beyond 'vision 2020'. *Economic and Political Weekly, 36*(11), 900–903.

Banerjee, S. B. (2007). Who sustains whose development? Sustainable development and the reinvention of nature. *Organization Studies, 24*(1), 143–180.

Baszanger, I., & Dodier, N. (2004). Ethnography: Relating the part to the whole. In D. Silverman (Ed.), *Qualitative research: Theory, method and practice* (pp. 9–35). London: Sage.

Baviskar, B. S. (1987). Cooperatives and rural development in India. *Current Anthropology, 28*(4), 564–565.

Bayly, S. (1999). *Caste, society and politics in India from the eighteenth century to the modern age.* Cambridge: Cambridge University Press.

Beck, U. (1992). *Risk society: Towards a new modernity.* London: Sage.

Beck, U. (1994). The reinvention of politics: Towards a theory of reflexive modernization. In U. Beck, A. Giddens, & S. Lash (Eds.), *Reflexive modernization: Politics, tradition and aesthetics in the modern social order* (pp. 1–55). Cambridge: Polity Press.

Beck, U. (1995). *Ecological politics in an age of risk.* Cambridge: Polity Press.

Beck, U. (1999). *World risk society.* Cambridge: Polity Press.

Beck, U. (2009). *World at risk.* Cambridge: Polity Press.

Beetham, D. (2013). *The legitimation of power.* London: Palgrave Macmillan.

Bernstein, S. (2004). *The elusive basis of legitimacy in global governance: Three conceptions, Working paper: GHC 04/2.* Hamilton: Institute on Globalization and the Human Condition.

Beteille, A. (1971). *Caste, class and power: Changing patterns of stratification in a Tanjore village.* London: University of California Press.

Bhagwati, J. (1998). The design of Indian development. In I. J. Ahluwalia & I. M. D. Little (Eds.), *India's economic reforms and development: Essays for Manmohan Singh* (pp. 23–39). Delhi: Oxford University Press.

Binz, C., Harris-Lovett, S., Kiparsky, M., Sedlak, D. L., & Truffer, B. (2016). The thorny road to technology legitimation – Institutional work for potable water reuse in California. *Technological Forecasting and Social Change, 103*, 249–263.

Bolognani, M. (2007). Islam, ethnography and politics: Methodological issues in researching amongst West Yorkshire Pakistanis in 2005. *Social Research Methodology, 10*(4), 279–293.

Bose, P. (2010). Women's reservation in legislatures: A defence. *Economic and Political Weekly, XLV*(14), 10–12.

Brass, P. R. (1990). *The politics of India since independence.* New York: Cambridge University Press.

Brown, J. M. (2008). *Mahatma Gandhi: The essential writings.* Oxford: Oxford University Press.

Bryld, E. (2001). Increasing participation in democratic institutions through decentralization: Empowering women and scheduled castes and tribes through panchayat raj in rural India. *Democratization, 8*(3), 149–172.

Buch, N. (2009). Reservation for women in panchayats: A sop in disguise? *Economic & Political Weekly, XLIV*(40), 8–10.

Butler, J. (2004). *Precarious life: The powers of mourning and violence.* London: Verso.

Buttel, F. H. (2005). The environmental and post-environmental politics of genetically modified crops and foods. *Environmental Politics, 14*(3), 309–323.

Carlsson, B., & Stanckiewicz, R. (1991). On the nature, function and composition of technological systems. *Journal of Evolutionary Economics, 1*, 93–118.

CESS. (2008). *Human development report 2007: Andhra Pradesh.* Hyderabad: Government of Andhra Pradesh.

Chandhoke, N., & Priyadarshi, P. (2009). Introduction. In N. Chandhoke & P. Priyadarshi (Eds.), *Contemporary India: Economy, society, politics* (pp. vii–xvii). New Delhi: Dorling Kindersley (India).

Chase, S. E. (1996). Personal vulnerability and interpretative authority in narrative research. In R. Josselson (Ed.), *Ethics and process in the narrative study of lives* (pp. 45–59). London: Sage.

Choudhary, B., & Gaur, K. (2010). *Bt cotton in India: A country profile.* New York: The International Service for the Acquisition of Agri-biotech Applications (ISAAA).

Choudhary, B., & Gaur, K. (2015). *Biotech cotton in India, 2002 to 2014.* New Delhi: The International Service for the Acquisition of Agri-biotech Applications (ISAAA).

Clark, I. (2007). Legitimacy in international or world society. In A. Hurrelmann, S. Schneider, & J. Steffek (Eds.), *Legitimacy in an age of global politics* (pp. 193–210). New York: Palgrave Macmillan.

Cohen, B. (2010). The historical context. In N. DeVotta (Ed.), *Understanding contemporary India*. London: Lynne Rienner Publishers.

Cohen, J. L., & Arato, A. (1992). *Civil society and political theory*. Cambridge: MIT Press.

Colman, G. P. (1968). Innovation and diffusion in agriculture. *Agricultural History, 42*(3), 173–188.

Conley, T., & Udry, C. (2001). Social learning through networks: The adoption of new agricultural technologies in Ghana. *American Journal of Agricultural Economics, 83*(3), 668–673.

Cooke, B. (2001). The social psychological limits of participation. In B. Cooke & U. Kothari (Eds.), *Participation: The new tyranny?* (pp. 102–121). London: Zed Books.

Corbridge, S., & Harriss, J. (2000). *Reinventing India: Liberalization, Hindu nationalism and popular democracy*. Cambridge: Polity Press.

Corbridge, S., Harriss, J., & Jeffrey, C. (2013). *India today: Economy, politics and society*. Cambridge: Polity Press.

Da Corta, L., & Venkateshwarlu, D. (1999). Unfree relations and the feminisation of agricultural labour in Andhra Pradesh. *Journal of Peasant Studies, 26*(2), 71–139.

Deaton, A., & Kozel, V. (2005). Data and dogma: The great Indian poverty debate. *The World Bank Research Observer, 20*(2), 177–199.

Deb, S. (2009). Public distribution of rice in Andhra Pradesh: Efficiency and reform options. *Economic and Political Weekly, XLIV*(51), 70–77.

Deshmukh, N. (2010). Cotton growers: Experience from Vidarbha. In R. S. Deshpande & S. Arora (Eds.), *Agrarian crisis and farmer suicides* (pp. 175–191). New Delhi: Sage.

Deshpande, R. (2004). Social movements in crisis? In R. Vora & S. Palshikar (Eds.), *Indian democracy: Meanings and practices* (pp. 379–409). New Delhi: Sage.

Deshpande, R. S., & Shah, K. (2010). Globalisation, agrarian crisis and farmers' suicides: Illusion and reality. In R. S. Deshpande & S. Arora (Eds.), *Agrarian crisis and farmer suicides* (pp. 118–148). New Delhi: Sage.

Desmond, E. (2014). *The legitimation of risk and democracy: A case study of Bt cotton in Andhra Pradesh, India*. Cork: University College Cork. Available at: https://cora.ucc.ie/handle/10468/1688/

Dev, S. M. (2012). *Small farmers in India: Challenges and opportunities*. Mumbai: Indira Gandhi Institute of Development Research.

Dev, S. M., & Rao, N. C. (2007). *Socioeconomic impact of Bt cotton.* Hyderabad: Centre for Economic and Social Studies (CESS).

Dinham, B. (2001). GM cotton – Farming by formula? *Biotechnology and Development Monitor, 44,* 7–9.

Dowd-Uribe, B. (2014). Engineering yields and inequality? How institutions and agro-ecology shape Bt cotton outcomes in Burkina Faso. *Geoforum, 53,* 161–171.

Elphinstone, M. (1866). *History of India: The Hindu Mohammedan periods.* London: John Murray.

Entine, J. (Ed.). (2006). *Let them eat precaution: How politics is undermining the genetic revolution in agriculture.* Washington, DC: American Enterprise Institute.

Epstein, S. T. (1973). *South India: Yesterday, today and tomorrow, Mysore villages revisited.* London: Macmillan Press.

Eyhorn, F. (2007). *Organic farming for sustainable livelihoods in developing countries? The case of cotton in India.* Zurich: vdf Hochschulverlag AG.

Faust, D., & Nagar, R. (2001). Politics of development in postcolonial India: English-medium education and social fracturing. *Economic and Political Weekly, 36*(30), 2878–2883.

Fischer, K. (2016). Why new crop technology is not scale neutral – A critique of the expectations for a crop-based African green revolution. *Research Policy, 45*(6), 1185–1194.

Forrester, D. B. (1970). Subregionalism in India: The case of Telangana. *Pacific Affairs, 43*(1), 5–21.

Forst, R. (2007). *The right to justification.* New York: Columbia University Press.

Forst, R. (2014a). *Justice, democracy and the right to justification.* London: Bloomsbury.

Forst, R. (2014b). *Justification and critique.* Cambridge: Polity Press.

Foucault, M. ([1976], 1994). Two lectures. In M. Kelly (Ed.), Critique and power: Recasting the Foucault/Habermas debate. London: Massachusetts Institute of Technology, pp. 17–46.

Foucault, M. (1977). *Discipline and punish.* London: Penguin.

Frankel, F. R. (1971). *India's green revolution: Economic gains and political costs.* Princeton: Princeton University Press.

Frankel, F. R. (2005). *India's political economy 1947–2004.* New Delhi: Oxford University Press.

Fraser, N. (2008). *Scales of justice: Reimagining political space in a globalizing world.* Cambridge: Polity Press.

Freeman, J. M. (1977). *Scarcity and opportunity in an Indian village.* California: Cummings Publishing Company.

Fricker, M. (2007). *Epistemic injustice: Power & the ethics of knowing.* Oxford: Oxford University Press.

Fukuda-Parr, S. (Ed.). (2007). *The gene revolution: GM crops and unequal development.* London: Earthscan.

Galab, S., & Rao, N. C. (2003). Women's self-help groups, poverty alleviation and empowerment. *Economic and Political Weekly, 38*(12/13), 1274–1283.

Galab, S., Revathi, E., & Reddy, P. P. (2009). Farmers' suicides and unfolding agrarian crisis in Andhra Pradesh. In D. N. Reddy & S. Mishra (Eds.), *Agrarian crisis in India* (pp. 164–198). New Delhi: Oxford University Press.

Gandhi, M. K. (1930, February 13). *Young India.*

Gandhi, M. K. ([1947], 2009). *India of my dreams.* Delhi: Rajpal.

Gandhi, M. K. (1948, January 18). *Harijan.*

Garikipati, S. (2009). Landless but not assetless: Female agricultural labour on the road to better status, evidence from India. *Journal of Peasant Studies, 36*(3), 517–545.

Gaurav, S., & Mishra, S. (2012). *To Bt or not to Bt? Risk and uncertainty considerations in technology assessment.* Mumbai: Indira Gandhi Institute of Development Research.

Geels, F. W., & Verhees, B. (2011). Cultural legitimacy and framing struggles in innovation journeys: A cultural-performative perspective and a case study of Dutch nuclear energy (1945–1986). *Technological Forecasting and Social Change, 78*, 910–930.

Geleta, E. (2013). The politics of identity and methodology in African development ethnography. *Qualitative Research, 0*(0), 1–16.

Gibson, C. (2012). Making redistributive direct democracy matter: Development and women's participation in the gram Sabhas of Kerala, India. *American Sociological Review, 77*(3), 409–434.

Giddens, A. (2003). *Runaway world: How globalization is reshaping our lives.* New York: Routledge.

Glover, D. (2010). Is *Bt* cotton a Pro-Poor Technology? A Review and Critique of the Empirical Record. *Journal of Agrarian Change, 10*(4), 482–509.

Goncalves, M. E. (2006). Risk and the governance of innovation in Europe: An introduction. *Technological Forecasting and Social Change, 73*, 1–12.

Goyal, L. C. (2015). *Accidental deaths and suicides in India, 2014.* Report from National Crime Records Bureau, New Delhi.

Griliches, Z. (1957). Hybrid corn: An exploration in the economics of technological change. *Econometrica, 25*(4), 501–522.

Gruère, G., & Sengupta, D. (2011). Bt cotton and farmer suicides in India: An evidence-based assessment. *Journal of Development Studies, 47*(2), 316–337.

Guillemin, M., & Gillam, L. (2004). Ethics, reflexivity, and "ethically important moments" in research. *Qualitative Inquiry, 10,* 261–280.

Gundimeda, S. (2009). Dalits, Praja Rajyam party and caste politics in Andhra Pradesh. *Economic and Political Weekly, 44*(21), 50–58.

Gupta, A. (2011). An evolving science-society contract in India: The search for legitimacy in anticipatory risk governance. *Food Policy, 36,* 736–741.

Gupta, S. P. (2002). *Report of the commission on India: Vision 2020.* New Delhi: Planning Commission, Government of India.

Gutierrez, A. P., Ponti, L., Herren, H. R., Baumgartner, J., & Kenmore, P. E. (2015). Deconstructing Indian cotton: Weather, yields and suicides. *Environmental Sciences Europe, 27*(12), 1–17.

Habermas, J. ([1973], 1976). *Legitimation crisis.* London: Heinemann Educational Books.

Habermas, J. (1996). *Between Facts and Norms: Contributions to a Discourse Theory of Law and Democracy.* Cambridge: Polity Press.

Habermas, J. (2001). *The postnational constellation: Political essays.* Cambridge: Polity Press.

Habermas, J. (2008). *Between naturalism and religion.* Cambridge, UK: Polity Press.

Hammersley, M., & Atkinson, P. (1995). *Ethnography: Principles in practice.* New York: Routledge.

Haragopal, G. (2010). The Telangana people's movement: The unfolding political culture. *Economic and Political Weekly, 45*(42), 51–60.

Harding, L., & Vidal, J. (2001, July 7). This is the path to disaster: Clare short is in the hot seat for funding GM crops in India. *The Guardian.*

Haunss, S. (2007). Challenging legitimacy: Repertoires of contention, political claims-making, and collective action frames. In A. Hurrelmann, S. Schneider, & J. Steffek (Eds.), *Legitimacy in an age of global politics* (pp. 156–172). New York: Palgrave Macmillan.

Hekkert, M. P., Suurs, R. A. A., Negro, S. O., Kuhlmann, S., & Smits, R. E. H. M. (2007). Functions of innovation systems: A new approach for analysing technological change. *Technological Forecasting and Social Change, 74,* 413–432.

Henrich, J. (2001). Cultural transmission and the diffusion of innovations: Adoption dynamics indicate that biased cultural transmission is the predominate force in behavioural change. *American Anthropologist, 103*(4), 992–1013.

Herring, R. J. (2008). Whose numbers count? Probing discrepant evidence on transgenic cotton in the Warangal district of India. *International Journal of Multiple Research Approaches, 2*(2), 145–159.

Herring, R. J., & Rao, N. C. (2012). On the 'failure of Bt cotton': Analyzing a decade of experience. *Economic and Political Weekly, XLVII* (18), 45–53.

Hindess, B., & Hirst, P. Q. (1975). *Pre-capitalist modes of production*. London: Routledge & Kegan Paul.

Hurrelmann, A., Schneider, S., & Steffek, J. (2007). Introduction: Legitimacy in an age of global politics. In A. Hurrelmann, S. Schneider, & J. Steffek (Eds.), *Legitimacy in an age of global politics* (pp. 1–16). New York: Palgrave Macmillan.

Inden, R. (1990). *Imagining India*. Oxford: Basil Blackwell.

Indrakanth, S. (1997). Coverage and leakages in PDS in Andhra Pradesh. *Economic and Political Weekly, 32*(19), 999–1001.

Iyengar, S., & Lalitha, N. (2007). GM cotton in Gujarat: General madness of genuine miracle? *Asian Biotechnology and Development Review, 9*(2), 45–81.

Iyer, K. G., & Arora, S. (2010). Indebtedness and farmers' suicides. In R. S. Deshpande & S. Arora (Eds.), *Agrarian crisis and farmer suicides* (pp. 264–291). New Delhi: Sage.

Jacobsen, K., & Landau, L. (2003). The dual imperative in refugee research: Some methodological and ethical considerations in social science research on forced migration. *Disasters, 27*(3), 185–206.

Jain, S. P. (2006). Panchayati raj in Andhra Pradesh yesterday, today and tomorrow. In S. B. Verma (Ed.), *Empowerment of the Panchayati raj institutions in India*. New Delhi: Sarup & Sons.

Jakimow, T. (2014). Gambling on livelihoods: Desire, hope and fear in agrarian Telangana. *Asian Journal of Social Science, 42*, 409–434.

Jangam, C. (2016). Dalit chronicles from the Telugu country. *Economic and Political Weekly, LI*(47), 25–29.

Jodhka, S. (2005). *Beyond 'crisis': Rethinking contemporary Punjab agriculture, Working paper* (Vol. 4). Hyderabad: CESS.

Johnson, C., Dowd, T. J., & Ridgeway, C. L. (2006). Legitimacy as a social process. *Annual Review of Sociology, 32*, 53–78.

Johnson, R. B., Onwuegbuzie, A. J., & Turner, L. A. (2007). Toward a definition of mixed methods research. *Journal of Mixed Methods Research, 1*(2), 112–133.

Jones, G. E. (1963). The diffusion of agricultural innovations. *Journal of Agricultural Economics, 15*, 387–409.

Kakar, S. (1995). *The colours of violence*. New Delhi: Penguin Books.

Kannabiran, K., Ramdas, S. R., Madhusudhan, N., Ashalatha, S., & Kumar, M. P. (2010). On the Telangana trail. *Economic and Political Weekly, XLV*(13), 69–81.

Karihaloo, J. L., & Kumar, P. A. (2009). *Bt cotton in India: A status report* (2nd ed.). New Delhi: Asia-Pacific Consortium on Agricultural Biotechnology and Asia-Pacific Association of Biotechnology Research Institutes.

Kathage, J., & Qaim, M. (2012). Economic impacts and impact dynamics of Bt (bacillus thuringiensis) cotton in India. *Proceedings of the National Academy of Sciences of the United States of America, 109*(29), 11652–11656.

Khare, R. S. (1983). Normative culture and kinship. In *Essays on Hindu categories, processes and perspectives.* Cambridge: Cambridge University Press.

Knorr-Cetina, K., & Mulkay, M. (Eds.). (1983). *Science observed: Perspectives on the social study of science.* London: Sage.

Kohli, A. (2009). *Democracy and development in India: From socialism to pro-business.* New Delhi: Oxford University Press.

Kolamkar, D. S. (2010). Report on the working of the Minimum Wages Act, 1948, for the year 2010. Government of India Ministry of Labour and Employment, Labour Bureau. http://labourbureau.nic.in/REP_MW_2010.pdf. Accessed 22 July 2013.

Kothari, R. (2005). *Rethinking democracy.* New Delhi: Orient Blackswan.

Krishna, V., Qaim, M., & Zilberman, D. (2016). Transgenic crops, production risk and biodiversity. *European Review of Agricultural Economics, 43*(1), 137–164.

Krishnaraj, M. (2006). Food security, agrarian crisis and rural livelihoods: Implications for women. *Economic and Political Weekly, 41*(52), 5376–5388.

Kulkarni, V. (2012). The making and unmaking of local democracy in an Indian village. *The Annals of the American Academy, 642,* 152–169.

Kumar, G. (2006). *Local democracy in India: interpreting decentralization.* New Delhi: Sage.

Kumar, P. (2009). *Panchayati raj institution in India.* New Delhi: Omega Publications.

Kumar, Y. V. A. (2010). *Seed bill 2010: An analytic view.* Centre for Sustainable Agriculture. http://agrariancrisis.in/wp-content/uploads/2012/08/SEED-BILL-2010-an-analytical-view.doc

Kumbamu, A. (2007). Discussion: Beyond agricultural deskilling and the spread of genetically modified cotton in Warangal. *Current Anthropology, 48*(6), 891–893.

Kuruganti, K. (2006). Biosafety and beyond: GM crops in India. *Economic and Political Weekly, 41*(40), 4245–4247.

Kuruganti, K. (2009). Bt Cotton and the Myth of Enhanced Yields. *Economic and Political Weekly, XLIV*(22), 29–33.

Latour, B. (1983). Give me a laboratory and I will raise the world. In K. Knorr-Cetina & M. Mulkay (Eds.), *Science observed: Perspectives on the social study of science* (pp. 141–170). London: Sage.

Laursen, L. (2012). Monsanto to face biopiracy charges in India. *Nature Biotechnology, 30*(1), 11.

Le Mons Walker, K. (2008). Neoliberalism on the ground in rural India: Predatory growth, agrarian crisis, internal colonization, and the intensification of class struggle. *Journal of Peasant Studies, 35*(4), 557–620.

LeCompte, M. D., & Schensul, J. J. (1999). *Book 1 – Ethnographer's toolkit: Designing and conducting ethnographic research*. London: Altamira Press.

Lerche, J. (2008). Transnational advocacy networks and affirmative action for Dalits in India. *Development and Change, 39*(2), 239–261.

Lidskog, R. (2008). Scientised citizens and democratised science. Re-assessing the expert-lay divide. *Journal of Risk Research, 11*(1–2), 69–86.

Locke, J. ([1690], 1967). *Two treatises of government*. Cambridge: Cambridge University Press.

Luhmann, N. (1993). *Risk: A sociological theory*. New York: Walter de Gruyter.

Mackenzie, C., McDowell, C., & Pittaway, E. (2007). Beyond 'do no harm': The challenge of constructing ethical relationships in refugee research. *Journal of Refugee Studies, 20*(2), 299–319.

Madan, V. (2007). *Co-operative movement in India*. New Delhi: Mittal Publications.

Mandelbaum, D. G. (1970a). *Society in India: Volume one continuity and change*. London: University of California Press.

Mandelbaum, D. G. (1970b). *Society in India: Volume two continuity and change*. Bombay: Popular Prakashan.

Maringanti, A. (2010). Telangana: Righting historical wrongs or getting the future right? *Economic & Political Weekly, XLV*(4), 33–38.

Markard, J., & Truffer, B. (2008). Technological innovation systems and the multi-level perspective: Towards an integrated framework. *Science Direct, 37*, 596–615.

Marx, K. (1853, June 25). The British rule in India. *New York Daily Tribune*.

Marx, K. ([1867], 2007). *Capital* (Vol. 1). New York: Cosimo.

McDowell, L. (1988). Coming in from the dark: Feminist research in geography. In J. Eyles (Ed.), *Research in human geography* (pp. 154–173). Oxford: Blackwell.

Mehta, J. L. (2005). *Advanced study in the history of modern India: 1707–1813*. New Delhi: New Dawn Press.

Mehta, P. B. (2007). The rise of judicial sovereignty. *Journal of Democracy, 18*(2), 70–83.

Melkote, R. S., Revathi, E., Lalita, K., Sajaya, K., & Suneetha, A. (2010). The movement for Telangana: Myth and reality. *Economic & Political Weekly, XLV*(2), 8–11.

Metcalf, B., & Metcalf, T. (2012). *A concise history of modern India.* New York: Cambridge University Press.

Michelutti, L. (2007). The vernacularization of democracy: Political participation and popular politics in North India. *Journal of the Royal Anthropological Institute, 13*, 639–656.

Miller, G., & Fox, K. J. (2004). Building bridges: The possibility of analytic dialogue between ethnography, conversation analysis and Foucault. In D. Silverman (Ed.), *Qualitative research: Theory, method and practice* (pp. 35–55). London: Sage.

Mishra, S. (2007). *Risks, farmers' suicides and agrarian crisis in India: Is there a way out?* Mumbai: Indira Gandhi Institute of Development Research.

Misra, S. S. (2009). No pesticides, no debts. *Down to Earth, 400*, 26–28.

Mohanty, M. (1986). Ideology and strategy of the communist movement in India. In T. Pantham & K. L. Deutsch (Eds.), *Political thought in modern India* (pp. 236–260). New Delhi: Sage.

Mohanty, B. (2005). 'we are like the living dead': Farmer suicides in Maharashtra, western India. *Journal of Peasant Studies, 32*(2), 243–276.

Moore, B., Jr. (1966). *Social origins of dictatorship and democracy: Lord and peasant in the making of the modern world.* Middlesex: Penguin University Books.

Morse, S., Bennett, R., & Ismael, Y. (2007). Inequality and GM crops: A case study of Bt cotton in India. *AgBioforum, 10*(1), 44–50.

Mulligan, S. (2007). Legitimacy and the practice of political judgement. In A. Hurrelmann, S. Schneider, & J. Steffek (Eds.), *Legitimacy in an age of global politics* (pp. 75–89). New York: Palgrave Macmillan.

Munshi, K. (2004). Social learning in a heterogenous population: Technology diffusion in the Indian green revolution. *Journal of Development Economics, 73*, 185–213.

Nair, R. (2010). Crop insurance in India: Changes and challenges. *Economic and Political Weekly, 45*(6), 19–22.

Nandi, S. (2010). Constructing the criminal: Politics of social imaginary of the "goonda". *Social Scientist, 38*(3/4), 37–54.

Nehru, J. (1958). *A bunch of old letters.* New Delhi: Asia Publishing House.

Nehru, J. (1964). *Jawaharlal Nehru's speeches volume 4.* Delhi: Government of India.

Omvedt, G. (1993). *Reinventing revolution: New social movements and the socialist tradition in India*. London: M.E. Sharpe.

Omvedt, G. (1994). *Dalits and the democratic revolution: Dr. Ambedkar and the Dalit movement in colonial India*. London: Sage.

Pantham, T. (1987). Habermas' practical discourse and Gandhi's *Satyagraha*. In B. Parekh & T. Pantham (Eds.), *Political discourse: Explorations in Indian and western political thought* (pp. 292–310). London: Sage.

Parthasarathi, G. (Ed.). (1989). *Jawaharlal Nehru: Letters to Chief Ministers Vol. 5 1958–64*. New Delhi: Jawaharlal Nehru Memorial Fund and Oxford University Press.

Patai, D. (1991). U.S. academics and third world women: Is ethical research possible? In S. B. Gluck & D. Patai (Eds.), *Women's words: The feminist practice of oral history*. London: Routledge.

Patnaik, U. (2007). Neoliberalism and rural poverty in India. *Economic and Political Weekly, 42*(30), 3132–3150.

Pattenden, J. (2005). Trickle-down solidarity, globalisation and dynamics of social transformation in a south Indian village. *Economic and Political Weekly, 40*(19), 1975–1985.

Pearson, M. (2006). 'Science,' representation and resistance: The Bt cotton debate in Andhra Pradesh, India. *The Geographical Journal, 172*(4), 306–317.

Pemunta, N. V. (2010). Intersubjectivity and power in ethnographic research. *Qualitative Research Journal, 10*(2), 3–19.

Pieterse, J. N. (2001). *Development theory: Deconstructions/reconstructions*. New Delhi: Vistaar.

Powis, B. (2003). Grass roots politics and 'second wave of decentralisation' in Andhra Pradesh. *Economic and Political Weekly, 38*(26), 2617–2622.

Prakash, C. S., & Conko, G. (2006). Agricultural biotechnology caught in a war of giants. In J. Entine (Ed.), *Let them eat precaution*. Washington, DC: American Enterprise Institute for Public Policy Research.

Pray, C. E., & Naseem, A. (2007). Supplying crop biotechnology to the poor: Opportunities and constraints. *Journal of Development Studies, 43*(1), 192–217.

Pur, K. A., & Moore, M. (2010). Ambiguous institutions: Traditional governance and local democracy in rural South India. *Journal of Development Studies, 46*(4), 603–623.

Qayum, A., & Sakkhari, K. (2005). *Bt cotton in Andhra Pradesh: A three-year assessment*. Hyderabad: Deccan Development Society.

Raina, R. S. (2006). Indo-US knowledge initiative: Need for public debate. *Economic and Political Weekly, 41*(17), 1622–1624.

Rajan, K. S. (2006). *Biocapital: The constitution of postgenomic life.* London: Duke University Press.

Ramakrishna, V. (1983). *Social reform in Andhra (1848–1919).* New Delhi: Vikas Publishing House.

Ramanjaneyulu, G. V., & Kuruganti, K. (2006). Bt cotton in India: Sustainable pest management? *Economic and Political Weekly, XLI*(7), 561–563.

Rangarajan, C. (2014). *Report of the expert group to review the methodology for measurement of poverty.* Delhi: Government of India Planning Commission.

Rao, V. M. (2009). Farmers' distress in a modernizing agriculture – The tragedy of the upwardly mobile: An overview. In D. N. Reddy & S. Mishra (Eds.), *Agrarian crisis in India* (pp. 109–125). New Delhi: Oxford University Press.

Rao, C. H. (2014). The new Telangana state: A perspective for inclusive and sustainable development. *Economic and Political Weekly, XLIX*(9), 10–13.

Rao, P. N., & Suri, K. C. (2006). Dimensions of agrarian distress in Andhra Pradesh. *Economic and Political Weekly, 41*(16), 1546–1552.

Rao, V., & Sanyal, P. (2010). Dignity through discourse: Poverty and the culture of deliberation in Indian village democracies. *The Annals of the American Academy, 629*, 146–172.

Ravi, T. (2015). Peasant struggles in Andhra Pradesh and Telangana: Reports from the field. *Review of Agrarian Studies, 5*(2), 110–125.

Rawls, J. ([1971], 1999). *A theory of justice: Revised edition.* Cambridge: Harvard University Press.

Ray, R. (1988). *The naxalites and their ideology.* Delhi: Oxford University Press.

Rayner, S. (2012). Uncomfortable knowledge: The social construction of ignorance in society and environmental policy discourses. *Economy and Society, 41*(1), 107–125.

Reddy, T. P. (2010). Distress and deceased in Andhra Pradesh. In R. S. Deshpande & S. Arora (Eds.), *Agrarian crisis and farmer suicides* (pp. 242–263). London: Sage.

Reddy, D. N., & Mishra, S. (2009). Agriculture in the reforms regime. In D. N. Reddy & S. Mishra (Eds.), *Agrarian crisis in India* (pp. 3–43). New Delhi: Oxford University Press.

Reddy, C. S., Jojaiah, K., Rao, N. V., & Narsaiah, I. (2012). Land and income inequalities in rural Andhra Pradesh. *The Marxist, XXVIII*, 50–74.

Renn, O. (2008). *Risk governance: Coping with uncertainty in a complex world.* London: Earthscan.

Revathi, E. (2009). 'Farmers' suicides in Andhra Pradesh: Issues and policy concerns. In S. M. Dev, C. Ravi, & M. Venkatanarayana (Eds.), *Human development in Andhra Pradesh: Experiences, issues and challenges*. Hyderabad: Centre of Economic and Social Studies.

Robinson, M. S. (1988). *Local politics: The law of the fishes: Development through political change in Medak district, Andhra Pradesh (South India)*. Delhi: Oxford University Press.

Roderick, R. (1986). *Habermas and the foundations of critical theory*. London: Macmillan.

Roosa, J. (2001). Passive revolution meets peasant revolution: Indian nationalism and the Telangana revolt. *Journal of Peasant Studies, 28*(4), 57–94.

Rousseau, J.-J. ([1762], 1973). *The social contract and discourses*. London: J.M. Dent & Sons.

Sainath, P. (2007). *The farm crisis: Why have over one lakh farmers killed themselves in the past decade?* New Delhi: Speaker's Lecture Series.

Sankaranarayanan, K., & Nalayini, P. (2015). Performance and behaviour of Bt cotton hybrids under sub-optimal rainfall situation. *Archives of Agronomy and Soil Science, 61*(8), 1179–1197.

Scanlon, T. M. (2012). Justification and legitimation: Comments on Sebastiano Maffettone's *Rawls: An introduction. Philosophy & Social Criticism, 38*(9), 887–892.

Scheper-Hughes, N. (1995). The primacy of the ethical: Propositions for a militant anthropology. *Current Anthropology, 36*(3), 409–440.

Schmitter, P., & Karl, T. L. (1991). What democracy is…and is not. *Journal of Democracy, 2*(3), Summer 1991, 75–88.

Scoones, I. (2005). *Science, agriculture and the politics of policy: The case of biotechnology in India*. Hyderabad: Orient Longman.

Scoones, I. (2008). Mobilizing against GM crops in India, South Africa and Brazil. *Journal of Agrarian Change, 8*(2 and 3), 315–344.

Sen, A. (1999). *Development as freedom*. Oxford: Oxford University Press.

Shah, E. (2005). Local and global elites join hands: Development and diffusion of Bt cotton technology in Gujarat. *Economic and Political Weekly, 40*(43), 4629–4639.

Sharma, D. C. (2004). Technologies for the people: A future in the making. *Futures, 36*, 733–744.

Sharma, S. (2010). Indian politics. In N. DeVotta (Ed.), *Understanding contemporary India* (pp. 67–94). London: Lynne Rienner.

Shiva, V. (1988). Reductionist science as epistemological violence. In A. Nandy (Ed.), *Science, hegemony and violence* (pp. 232–257). Delhi: Oxford University Press.

Shiva, V. (1991). *The violence of the green revolution: Third world agriculture, ecology and politics.* London: Zed.

Shiva, V., & Jafri, A. (1998). *Seeds of suicide: The ecological and human costs of globalization of agriculture.* New Delhi: Research Foundation for Science, Technology, Ecology.

Shiva, V., Emani, A., & Jafri, A. (1999). Globalisation and threat to seed security: Case of transgenic cotton trials in India. *Economic and Political Weekly, 34*(10–11), 601–613.

Shurmer-Smith, P. (2000). *India: Globalization and change.* London: Arnold.

Shylendra, H. S. (2006). Microfinance institutions in Andhra Pradesh: Crisis and diagnosis. *Economic and Political Weekly, 41*(20), 1959–1963.

Simon-Kumar, R. (2014). Sexual violence in India: The discourses of rape and the discourses of justice. *Indian Journal of Gender Studies, 21*(3), 451–460.

Singh, M. P., & Verney, D. V. (2003). Challenges to India's centralised parliamentary federalism. *Publius, 33*(4), 1–20.

Smale, M., Zambrano, P., & Cartel, M. (2006). Bales and balance: A review of the methods used to assess the economic impact of Bt cotton on farmers in developing economies. *AgBioforum, 9*(3), 195–212.

Smith, J. M. (2004). *Seeds of deception: Exposing corporate and government lies about the safety of genetically engineered food.* Devon: Green Books.

Smith, J. M. (2007). *Genetic roulette: The documented health risks of genetically engineered foods.* Vermont: Chelsea Green.

Sridhar, V. (2006). Why do farmers commit suicide? The case of Andhra Pradesh. *Economic and Political Weekly, 41*(16), 1559–1565.

Srikrishna, B. N., Duggal, V. K., Singh, R., Shariff, A., & Kaur, R. (2010). *Committee for consultations on the situation in Andhra Pradesh.* New Delhi: Government of India.

Srinivas, M. N. (1966). *Social change in modern India.* New Delhi: Orient Longman.

Srinivas, M. N. (1987). *The dominant caste and other essays.* Delhi: Oxford University Press.

Srinivas, M. N. (1991). On living in a revolution. *Economic and Political Weekly, 26*(13), 834–836.

Srinivas, M. N. (2003). *Religion and society among the Coorgs of South India.* New Delhi: Oxford University Press.

Srinivasulu, K. (2002). *Caste, class and social articulation in Andhra Pradesh: Mapping differential regional trajectories.* London: Overseas Development Institute.

Srinivasulu, K., & Sarangi, P. (1999). Political realignments in post-NTR Andhra Pradesh. *Economic and Political Weekly, 34*(34/35), 2449–2458.

Srivastava, S. K., & Kolady, D. (2016). Agricultural biotechnology and crop productivity: Macro-level evidences on contribution of Bt cotton in India. *Current Science, 110*(3), 311–319.

Stone, G. D. (2007). Agricultural deskilling and the spread of genetically modified cotton in Warangal. *Current Anthropology, 48*(1), 67–103.

Stone, G.D. (2011a). Contradictions in the last mile: Suicide, culture, and e-agriculture in rural India. *Science, Technology & Human Values, 36*(6), 759–790.

Stone, G. D. (2011b). Field *versus* farm in Warangal: Bt cotton, higher yields, and larger questions. *World Development, 39*(3), 387–398.

Strydom, P. (2002). *Risk, environment and society.* Buckingham: Open University Press.

Strydom, P. (2008). Risk communication: World creation through collective learning under complex contingent conditions. *Journal of Risk Research, 11*(1–2), 5–22.

Subramaniam, A., & Qaim, M. (2010). The impact of Bt cotton on poor households in rural India. *Journal of Development Studies, 46*(2), 295–311.

Sundarayya, P. (1972). *Telangana people's struggle and its lessons.* Calcutta: Desraj Chadha on behalf of the Communist Party (Marxist).

Swaminathan, M. S. (2011). *In search of biohappiness: Biodiversity and food, health and livelihood security.* London: World Scientific.

Szerszynski, B. (1996). On knowing what to do: Environmentalism and the modern problematic. In S. Lash, B. Szerszynksi, & B. Wynne (Eds.), *Risk, environment & modernity: Towards a new ecology.* London: Sage.

Tesoriero, F. (2005). Strengthening communities through women's self help groups in South India. *Community Development Journal, 41*(3), 321–333.

Townsend, R. M. (1994). Risk and insurance in village India. *Econometrica, 62*(3), 539–591.

Vaikuntham, Y. (2004). *Studies in socio-economic and political history: Hyderabad state.* Hyderabad: Karshak Art Printers.

Vakulabharanam, V. (2004). Agricultural growth and irrigation in Telangana: A review of evidence. *Economic and Political Weekly, 39*(13), 1421–1426.

Varshney, A. (1998). *Democracy, development and the countryside: Urban-rural struggles in India*. New York: Cambridge University Press.

Vasavi, A. R. (2010). Contextualising the agrarian suicides. In R. S. Deshpande & S. Arora (Eds.), *Agrarian crisis and farmer suicides* (pp. 70–85). New Delhi: Sage.

Vasavi, A. R. (2012). *Shadow space: Suicides and the predicament of rural India*. New Delhi: Three Essays Collective.

Venkat, V. (2015, June 21). Bt cotton responsible for suicides in rain-fed areas, says study. *The Hindu*.

Venkateshwarlu, D., & Srinivas, K. (2000). *Debt and deep well: Status of small and marginal farmers in Warangal district*. Hyderabad: Care AP; ICNGO Programme AP; MARI.

Verma, P. (2005). Female infanticide – Not a practice of the past. *Off Our Backs, 35*(3/4), 28–31.

Visvanathan, S., & Parmar, C. (2002). A biotechnology story: Notes from India. *Economic and Political Weekly, 37*(27), 2714–2724.

Vyas, V. S., & Singh, S. (2006). Crop insurance in India: Scope for improvement. *Economic and Political Weekly, 41*(43/44), 4585–4594.

Wade, R. (1994). *Village republics: Economic conditions for collective action in south India*. San Francisco: ICS Press.

Walker, T. S., & Ryan, J. G. (1990). *Village and household economies in India's semi-arid tropics*. London: The Johns Hopkins University Press.

Weber, M. ([1968], 1978). *Economy and society: An outline of interpretative sociology* (Vol. 1). London: University of California Press.

Weiner, M. (2001). The struggle for equality: Caste in Indian politics. In A. Kohli (Ed.), *The success of India's democracy* (pp. 193–225). New Delhi: Cambridge University Press.

Weis, T. (2007). *The global food economy: The battle for the future of farming*. New York: Palgrave Macmillan.

Wolf, D. L. (1996). Situating feminist dilemmas in fieldwork. In D. L. Wolf (Ed.), *Feminist dilemmas in fieldwork*. Oxford: Westview Press.

Wynne, B. (2001). Creating public alienation: Expert cultures of risk and ethics on GMOs. *Science as Culture, 10*(4), 445–481.

Young, I. M. (2000). *Inclusion and democracy*. New York: Oxford University Press.

Web-Sites and Newspaper Articles

Chapter One

Web-Sites

Gene Watch. UK: 'Worldwide commercial growing' 2015. http://www.gene-watch.org/sub-532326. Accessed 24 Mar 2017.

GMO Compass. USA 2013: 'No reversal of trend – Farmers stick to their varieties of GM crops', 21/5/2014. http://www.gmo-compass.org/eng/agri_biotech-nology/gmo_planting/506.usa_cultivation_gm_plants_2013.html. Accessed 24 Mar 2017.

Jishnu, L. (2010). Bt cotton: Monsanto is back in courts over royalty, 1/4/2010. http://business.rediff.com/column/2010/apr/01/guest-bt-cotton-monsanto-is-back-in-courts-over-royalty.htm. Accessed 25 Mar 2017.

Kesireddy, R. (2015). Telangana set for GM trials by seed giants, 13/11/2015. http://articles.economictimes.indiatimes.com/2015-11-13/news/68252428_1_field-trials-gm-crops-genetic-engineering-appraisal-committee. Accessed 25 Mar 2017.

Mathur, S. (2010). Countrywide protest against Bt brinjal. *The Times of India*, 31/1/2010. http://timesofindia.indiatimes.com/india/Countrywide-protest-against-Bt-brinjal/articleshow/5518257.cms. Accessed 24 Mar 2017.

Monsanto Company 2006 Annual Report. (2006). It all starts today. http://www.monsanto.com/investors/documents/pubs/2006/2006annualreport.pdf. Accessed 24 Mar 2017.

Salve, P. (2014). How many farmers does India really have? *Hindustan Times*, 11/8/2014. http://www.hindustantimes.com/india/how-many-farmers-does-india-really-have/story-431phtct5O9xZSjEr6HODJ.html. Accessed 24 Mar 2017.

Sainath, P. (2013). Over 2,000 fewer farmers every day. *The Hindu*, 2/5/2013. http://www.thehindu.com/opinion/columns/sainath/over-2000-fewer-farmers-every-day/article4674190.ece. Accessed 24 Mar 2017.

Sustainable Pulse. GM crops now banned in 38 countries worldwide – Sustainable pulse research, 22/10/2015. http://sustainablepulse.com/2015/10/22/gm-crops-now-banned-in-36-countries-worldwide-sustainable-pulse-research/#.VqTFr5qLSt9. Accessed 24 Mar 2017.

Chapter Two

Newspapers

Shrinivasan, R. 55% of India's population poor. *Times of India,* 16/7/2010.

Web-Sites

Census of India. (2011a). Sex ratio in India. http://www.census2011.co.in/ sexratio.php. Accessed 25 Mar 2017.

Census of India. (2011a). Provisional population totals: Rural-urban distribution: Figures at a glance. http://www.censusindia.gov.in/2011-prov-results/ paper2/data_files/india/paper2_at_a_glance.pdf. Accessed 25 Mar 2017.

Census of India. (2011b). Literacy in India. http://www.census2011.co.in/literacy.php. Accessed 25 Mar 2017.

Chapter Three

Newspapers

DC Correspondent, New Delhi: 'Seven 2G firms get ED notices for laundering'. *Deccan Chronicle,* 24/11/2010.

Srinivas, P. (2010). Jobs fuel liquor sales. *Deccan Chronicle,* 20/10/2010.

Srivastava, R. AP's poor groaning under debt trap. *Times of India,* 12/10/2010.

Times News Network: '2 killed as police open fire on protesters in Srikakulam'. *Times of India,* 15/7/2010.

Web-Sites

Approach to the 12th Five Year Plan: Andhra Pradesh, January 2013, Centre for Economic and Social Studies, Hyderabad. http://www.cess.ac.in/cesshome/ pdf/draft_approach_to_12th_plan_for_discussion.pdf. Accessed 26 Mar 2017.

ASTM. (2011). August 9th declared as "Monsanto quit India!" day, 9/8/2011. http://astm.lu/august-9th-declared-as-monsanto-quit-india-day/. Accessed 26 Mar 2017.

Constituent Assembly of India, Volume II 22/1/1947. http://parliamentofindia.nic.in/ls/debates/vol2p3.htm. Accessed 26 Mar 2017.

Environment Support Group Trust. Say no to Brinjal, say no to release of genetically modified crops in India: Participate in the public consultation to be held on Bt Brinjal by Shri. Jairam Ramesh. 6/2/2010. http://www.esgindia.org/campaigns/press/say-no-bt-brinjal-say-no-release-genetic.html. Accessed 27 Mar 2010.

India's GDP growth slumps to 6.5 per cent in 2011–12. *India Today,* 31/5/2012. http://indiatoday.intoday.in/story/gdp-growth-sharply-down-at-6.5-per-cent-in-2011-12/1/198325.html. Accessed 26 Mar 2017.

Jishnu, L. Bt cotton: Monsanto is back in courts over royalty. *Rediff.com,* 1/4/2010. http://business.rediff.com/column/2010/apr/01/guest-bt-cotton-monsanto-is-back-in-courts-over-royalty.htm. Accessed 27 Mar 2017.

Kurunganti, K. Mass protests against GM crops in India. *Science in Society Archive,* 30/4/2008. http://www.i-sis.org.uk/gmProtestsIndia.php. Accessed 3 Apr 2017.

Ministry of Environment and Forests, Decision on Commercialisation of Bt Brinjal. http://webcache.googleusercontent.com/search?q=cache:http://www.moef.nic.in/downloads/public-information/minister_REPORT.pdf&gws_rd=cr&ei=0KjGWKudLtOogAbDmqjAAQ. Accessed 27 Mar 2010.

Monsanto told to shell out fine. *Deccan Chronicle,* 1/2/2006. http://www.global-sisterhood-network.org/content/view/705/76/. Accessed 27 Mar 2017.

Monsanto, AP govt cross swords over royalty payment. *The Hindu,* 22/4/2006. http://www.thehindubusinessline.com/todays-paper/tp-agri-biz-and-commodity/monsanto-ap-govt-cross-swords-over-royalty-payment/article1731571.ece?ref=archive. Accessed 27 Mar 2017.

Nandy, A. (1996). Sustaining the faith. *India Today,* 31/8/1996. http://indiatoday.intoday.in/story/indian-democracy-today-is-stronger-in-the-minds-of-people-than-it-ever-was/1/283011.html. Accessed 26 Mar 2017.

Parliament urged to pass BRAI, Seeds Bill. *The Hindu,* 7/8/2011. http://www.thehindu.com/todays-paper/tp-national/parliament-urged-to-pass-brai-seeds-bill/article2332299.ece. Accessed 3 Apr 2017.

Press Information Bureau. Government of India: Re-inclusion of cotton seed as Essential Commodity under the Essential Commodities Act, 1955, Release ID 54255, 19/11/2009. http://pib.nic.in/newsite/erelease.aspx?relid=54255. Accessed 27 Mar 2017.

Rao, A. S. 95 per cent people below poverty line in Andhra Pradesh. *India Today*, 17/12/2011. http://indiatoday.intoday.in/story/95-per-cent-below-poverty-line-andhra-pradesh/1/164651.html. Accessed 26 Mar 2017.

RT Question More Live. (2015). World stands up against Monsanto: Over 400 cities protest GMOs, 24/5/2015. https://www.rt.com/news/261573-monsanto-global-protests-gmo/. Accessed 26 Mar 2017.

Scientists, farmers fast to protest Bt brinjal. *The Hindu*, 30/1/2010. http://www.thehindu.com/sci-tech/agriculture/Scientists-farmers-fast-to-protest-Bt-Brinjal/article16840509.ece. Accessed 27 Mar 2017.

Shukla, A. (2010). First official estimate: An NGO for every 400 people in India. *The Indian Express*, 7/7/2010. http://www.indianexpress.com/news/first-official-estimate-an-ngo-for-every-400-people-in-india/643302/. Accessed 26 Mar 2017.

Sonwalkar, P. (2013). NREGA benefits are mixed: Oxford study. *Hindustan Times*, 2/11/2013. http://www.hindustantimes.com/india/nrega-benefits-are-mixed-oxford-study/story-ofHsUDTESGXETmF6rJDjJM.html. Accessed 26 Mar 2017.

The Andhra Pradesh Gazette: Part IV-B Extraordinary, No. 36, Andhra Pradesh Acts, Ordinances and Regulations, etc., 16/8/2007. http://faolex.fao.org/docs/pdf/ind119055.pdf. Accessed 27 Mar 2017.

The Constitution (Seventy-Third Amendment Act) 1992. http://indiacode.nic.in/coiweb/amend/amend73.htm. Accessed 26 Mar 2017.

Venkateshwarlu, K. (2013). Andhra Pradesh leads in bringing down poverty. *The Hindu*, 25/7/2013. http://www.thehindu.com/news/national/andhra-pradesh/andhra-pradesh-leads-in-bringing-down-poverty/article4952012.ece. Accessed 26 Mar 2017.

Chapter Four

Newspaper Articles

DC Correspondent: 'T-demand takes one more life'. *Deccan Chronicle*, 16/8/2010.

DC Correspondent: 'T activists gherao ministers'. *Deccan Chronicle*, 9/1/2011 9/1/2011.

DC Correspondents: 'RTC [bus service] faces the brunt as bandh hits T areas'. *Deccan Chronicle*, 22/2/2011.

DC Correspondents: 'T men derail traffic, have a ball on tracks'. *Deccan Chronicle*, 2/3/2011.

In Naxal Land Govt Faces Trust Deficit. *Times of India*, 28/9/2010.

Mahapatra, D. Skewed development fuels Naxalism: SC [Supreme Court]. *Times of India*, 21/7/2010.

Mitta, M. (2011). The case for Telangana. *Times of India*, 10/1/2011.

Prasad, B. K. Macro plan to rein in micro finance institutions. *Deccan Chronicle*, 14/10/2010.

Srivastava, R. MFIs no better than moneylenders. *Times of India*, 14/10/2010.

Srivastava, R.. What kills borrowers? *Times of India*, 17/10/2010.

Times News Network (Warangal): 'Daily wage worker hangs himself for T cause'. *Times of India*, 17/10/2010.

Times News Network (Warangal): 'Student sets self on fire over 'T delay''. *Times of India*, 12/2/2011.

Times News Network: 'Auto driver ends life for T in Karimnagar'. *Times of India*, 29/1/2011.

Times News Network: 'Bandh near total in Telangana'. *Times of India*, 8/1/2011.

Times News Network: 'CM [Chief Minister] faces T [Telangana] women's wrath'. *Times of India*, 11/2/2011.

Times News Network: 'KCR [local politician] ditches cadre at relay hunger strike'. *Times of India*, 22/1/2011.

Times News Network: 'Maoist terror resurfaces in T [Telangana]: Two shot dead, 3 Abducted by Naxals'. *Times of India*, 3/12/2010.

Times News Network: 'Maoists torch 14 vehicles'. *Times of India*, 22/12/2010.

Times News Network: 'Naxal leader shot dead in cold blood, says rights forum'. *Times of India*, 28/1/2011.

Times News Network: 'Naxals release three tribals'. *Times of India*, 7/12/2010.

Times News Network: 'Trouble erupts in OU [Osmania University] again'. *Times of India*, 25/1/2011.

Times News Network: 'OU [Osmania University] student takes poison during [Telangana] March'. *Times of India*, 11/3/2011.

Unattributed (New Delhi). CM promises ordinance to curb MFIs. *Times of India*, 13/10/2010.

Web-Sites

Agriculture in Warangal, India. http://www.sourcewatch.org/index.php/Agriculture_in_Warangal,_India. Accessed 27 Mar 2017.

Census 2011. Warangal city census 2011 data. http://www.census2011.co.in/census/city/398-warangal.html. Accessed 27 Mar 2017.

Census 2011. Warangal district: Census 2011 data. http://www.census2011. co.in/census/district/126-warangal.html. Accessed 27 Mar 2017.

Census of India 2011: Provisional population totals: Data on urban and rural areas. Figures at a glance: Andhra Pradesh http://www.censusindia.gov. in/2011-prov-results/paper2/data_files/AP/4-fig-6.pdf. Accessed 27 Mar 2017.

Centre for Sustainable Agriculture: Caring for those who feed the nation. http:// csa-india.org/what-we-do/npm/. Accessed 27 Mar 2017.

Findthedata.org: Andhra Pradesh. http://states-of-india.findthedata.org/q/1/4243/ How-many-districts-are-there-in-Andhra-Pradesh. Accessed 27 Mar 2017.

Hongthong, P. (2004). Bitter-sweet of GM crops. *Yale Global Online,* 7/1/2004. http://yaleglobal.yale.edu/content/bitter-sweet-gm-crops. Accessed 27 Mar 2017.

Jebaraj, P. (2012). Bhoodan land at centre of unparalleled scam, says Ramesh, 21/9/2012. http://www.thehindu.com/news/national/bhoodan-land-at-centre-of-unparalleled-scam-says-ramesh/article3919255.ece. Accessed 27 Mar 2017.

Kohli, K. (2005). Centre's no to Bt cotton in AP. *India Together,* 1/5/2005. http://www.indiatogether.org/2005/may/agr-apcotton.htm. Accessed 27 Mar 2017.

Monsanto Bt cotton fails in Warangal. *The Hindu,* 16/10/2004. http://www. hindu.com/2004/10/16/stories/2004101604770500.htm. Accessed 27 Mar 2017.

Non Pesticide Management in Andhra Pradesh, India. ftp://ftp.fao.org/sd/sda/ sdar/sard/GP%20updates/pest_management_India. Accessed 29 Mar 2017.

Reinventing Telangana, The way forward: Socio-economic outlook 2016. Government of Telangana planning department http://sakshieducation.com/ Budget/TS/TS-Socio-Economic-2016.pdf. Accessed 27 Mar 2017.

Report of the commission on farmers' welfare, Government of Andhra Pradesh. http://www.macroscan.org/pol/apr05/pdf/Full_Report_Commission_ Farmer_AP.pdf. Accessed 27 Mar 2017.

Revenue Disaster Management. Government of Andhra Pradesh: Vulnerability of the state. http://disastermanagement.ap.gov.in/historyofdisasters.aspx. Accessed 27 Mar 2017.

Telangana gets 21 new districts. *The Hindu,* 1/12/2016. http://www.thehindu. com/news/national/telangana/Telangana-gets-21-new-districts/arti-cle15479100.ece. Accessed 27 Mar 2017.

Telangana state portal: State profile. http://www.telangana.gov.in/about/state-profile. Accessed 27 Mar 2017.

The great Telangana shudder: 72 MLAs resign and more to join. *One India,* 4/7/2011. http://www.oneindia.com/2011/07/04/telangana-resignations-72-mla-s-resign-more-to-join-aid0113.html. Accessed 27 Mar 2017.

Chapter Five

Newspaper Articles

DC Correspondent: 'Farmer ends life over debt'. *Deccan Chronicle,* 18/12/2010.

DC Correspondent: 'Minister: 49 farmers died in December'. *Deccan Chronicle,* 27/12/2010.

DC Correspondents: '10 farmers die due to rain damage'. *Deccan Chronicle,* 5/1/2011.

DC Correspondents: 'Five farmers die over debt, crop losses'. *Deccan Chronicle,* 26/12/2010.

DC Correspondents: 'Heavy rains take a toll on crops'. *Deccan Chronicle,* 9/12/2010.

Times News Network: 'Cotton farmers go berserk, beat up traders'. *Times of India,* 30/10/2010.

Times News Network: 'Cotton growers left in the lurch'. *Times of India,* 7/1/2011.

Times News Network: 'Crop losses claim lives of 10 more ryots [cultivators]'. *Times of India,* 30/12/2010.

Times News Network: 'Mounting debts drive farmer to suicide'. *Times of India,* 8/10/2010.

Times News Network: 'Seven more farmers die of crop-loss shock'. *Times of India,* 20/12/2010.

Times News Network: 'Three more farmers die of shock'. *Times of India,* 18/12/2010.

Times News Network: 'Unable to bear crop loss, 3 farmers die'. *Times of India,* 10/12/2010.

Times News Network: 'Upset over crop loss, woman farmer commits suicide'. *Times of India,* 9/12/2010.

Chapter Six

Newspaper Articles

Chiru: Suicides increased after AP announced relief. *Deccan Chronicle,* 27/12/2010.

Web-Sites

Bitter Seeds: An examination of the debate surrounding biotechnology and the future of farming. *ITVS*, 1/4/2013. http://www.itvs.org/films/bitter-seeds. Accessed 30 Mar 2017.

Press Information Bureau, Government of India, Ministry of Finance. (2014). Agricultural debt waiver and debt relief scheme (ADWDRS), 2008. http://pib.nic.in/newsite/PrintRelease.aspx?relid=104122. Accessed 30 Mar 2017.

YSR credited with loan waiver scheme. *The Hindu,* 17/7/2008. http://www.thehindu.com/todays-paper/tp-national/tp-andhrapradesh/YSR-credited-with-loan-waiver-scheme/article15262124.ece. Accessed 30 Mar 2017.

Chapter Seven

Web-Sites

Government of Andhra Pradesh: Panchayat Raj & Rural Development (R.D.II) Dept. http://www.rd.ap.gov.in/EGS/SA_Rules_170408_final.pdf. Accessed 30 Mar 2017.

GM Watch. Women march for total ban on Bt cotton, 14/4/2012. http://www.gmwatch.org/index.php/news/archive/2012/13848-women-march-for-total-ban-on-bt-cotton. Accessed 30 Mar 2017.

Chapter Eight

Web-Sites

Government of Telangana Planning Department. Reinventing Telangana, The way forward: Socio economic outlook 2016. www.telangana.gov.in/PDFDocuments/Socio-Economic-Outlook-2016.pdf. Accessed 31 Mar 2017.

GM Watch. Narendra Modi: Monsanto's man in India, 28/3/2015. http://www.gmwatch.org/index.php/news/archive/2015-articles/16038. Accessed 2 Apr 2017.

Kesireddy, R. R. (2015). Telangana set for GM trials by seed giants. *The Economic Times*, 13/11/2015. http://economictimes.indiatimes.com/news/economy/agriculture/telangana-set-for-gm-trials-by-seed-giants/articleshow/49765702.cms?intenttarget=no. Accessed 2 Apr 2017.

Index[1]

---

[1]Note: Page numbers followed by 'n' refer to notes.

© The Author(s) 2018
E.L. Desmond, *Legitimation in a World at Risk*,
https://doi.org/10.1007/978-981-10-6065-6